Roman Gr. Maev and Volf Leshchynsky

Introduction to Low Pressure Gas Dynamic Spray

Related Titles

Pawlowski, L.

The Science and Engineering of Thermal Spray Coatings 2e

2007. Hardcover

ISBN: 978-0-471-49049-4

Lavernia, E.J. and Wue, Y.

Spray Atomization & Deposition

1996. Hardcover

ISBN: 978-0-471-95477-4

Roman Gr. Maev and Volf Leshchynsky

Introduction to Low Pressure Gas Dynamic Spray

Physics & Technology

WILEY-VCH Verlag GmbH & Co. KGaA

The Authors

Prof. Roman Maev
NSERC Industrial Research Chair
University of Windsor
401, Sunset Avenue
Windsor, ON N9B 3P4
Canada

Dr. Volf Leshchynsky
Dept. of Physics
University of Windsor
401 Sunset Ave.
Windsor, ON N9B 3P4
Canada

Library of Congress Card No.:
applied for

British Library Cataloguing-in-Publication Data
A catalogue record for this book is available from the British Library.

Bibliographic information published by the Deutsche Nationalbibliothek
The Deutsche Nationalbibliothek lists this publication in the Deutsche Nationalbibliografie; detailed bibliographic data are available in the Internet at <http://dnb.d-nb.de>.

Composition Da-TeX, Gerd Blumenstein, Leipzig
Printing Strauss GmbH, Mörlenbach
Binding Litges& Dopf Buchbinderei GmbH, Heppenheim

Printed in the Federal Republic of Germany
Printed on acid-free paper

ISBN: 978-3-527-40659-3

Contents

Introduction to Low Pressure Gas Dynamic Spray Roman Gr. Maev and Volf Leshchynsky

ISBN: 978-3-527-40659-3

Preface

After many years of intensive research and development, as well as numerous trials, cold spraying or gas dynamic spraying (GDS) has completed the transition from an emerging process to a commercially viable method for coating various high performance machine components within a manufacturing environment. To take advantage of the performance capabilities of the GDS process, such leading industrial companies as DaimlerChrysler, Ford, and Alcoa are gradually implementing this new coating technique into their engineering processes. An additional boost to the industrial application of GDS technology was provided as a result of the development of a portable low-pressure GDS machine by Centerline Windsor Ltd. (Windsor, ON, Canada).

As portable GDS machines have considerably expanded the application possibilities of the low-pressure coating deposition, it is beneficial to examine the main trends and developments behind this technique. Further, it is advantageous to present a comprehensive assessment of low-pressure GDS technology as it fits within the framework of thermal spraying manufacturing processes. This book analyzes the physical background of low-pressure GDS, the metallurgical and mechanical principles involved, and presents a detailed description of low-pressure GDS practices and applications. It is shown that the successful development of the described technology results from the proper application of metallurgical and powder material. This may be achieved through experimental and theoretical methods of analysis. Now, although low- and high-pressure GDS are competing processes, in-depth comparison of these technologies reveals their common background and indeed overlapping opportunities for application. As such, readers who are new to the field of GDS are encouraged to supplement their general knowledge through one of the many excellent sources noted in the text. In this book, emphasis is placed on the proper development of the low-pressure GDS coatings and their structure and properties, which are of paramount importance to the success of GDS. Various examples of low-pressure GDS coatings and deposition methodologies are also presented.

Introduction to Low Pressure Gas Dynamic Spray Roman Gr. Maev and Volf Leshchynsky

ISBN: 978-3-527-40659-3

The primary objective of this text is to emphasize the importance of a combined metallurgy/powder technology approach to the development of the GDS process. The relationship between the microstructure, composition, and the mechanical properties of the sprayed material, as well as the role of deformation, powder densification, and consolidation in enhancing these properties are discussed in depth. This interdisciplinary approach is a major catalyst in the successful design of low-pressure GDS processes. To facilitate this approach, we illustrate two GDS process design procedures: (i) powder shock consolidation analysis of the GDS process, and (ii) defining the structure–properties relationship for coating characterization. These procedures are extremely important in the rapid development of low-pressure GDS, as well as the assurance of high quality coatings.

The authors would like to express our appreciation to the many individuals within government agencies, companies, and universities who have pushed low-pressure GDS technology forward during the past 10 years. Their projects and public dissemination of the technical results contributed immeasurably to the emergence of low-pressure GDS technologies as a strong and growing manufacturing process. We would like to express our kind appreciation to Professor A. Papyrin, Dr. D. Helfritch, Dr. T. Van Steenkiste, Dr. Emil Strumban, Mr. Michael Beneteau, and Mr. Wally Birtch for their comments as well as for providing some of the data used in the book. Thanks are extended to Mr. Wesley Arthur for his assistance in editing, proofreading, and assembling this manuscript, and to Mr. Jeffry Sadler for converting it to LaTex. Obviously we would like to thank Christoff and Ester from Wiley and Sons for inviting us to publish this book and for their patience in working with us. Finally, we acknowledge the support, tolerance, and understanding of our families throughout this endeavor.

Windsor, Canada May 2007 Roman Maev, Volf Leshchynsky

1 Introduction

1.1 General Description

Cold gas dynamic spray (GDS) is a rapidly emerging coating technology, in which spray particles in a solid state are deposited on a substrate via high-velocity impact, at temperatures lower than the melting point of the powder material (Papyrin et al. 2006). As shown by McCune et al. (2000), the methodology for generating surface coatings from metallic particles in solid form was first patented by Rocheville (1963). This process employs a supersonic nozzle into which solid particles of feedstock material are introduced and accelerated toward a substrate, either forming a thin surface coating, or being directly embedded as isolated particles in small depressions on the surface. An alternative technique for producing thick deposits of various metals from "cold" jets was reported by Alkhimov et al. (1990). In this method, solid particles ranging in size from 10 to 50 μm are introduced by means of a pressurized powder feeder into the high-pressure, high-temperature chamber of a converging–diverging (de Laval) nozzle and are subsequently accelerated into a supersonic stream by the gas (Fig. 1.1(a)). Here, as the gas has been electrically preheated prior to its introduction into the nozzle, the term "cold spraying" appears to contradict with the nature of this deposition process. For this reason this process is more appropriately referred to as "Gas Dynamic Spray."

In deposition, spray materials experience only minor changes in microstructure and little oxidation of decomposition. Most metals such as Cu, Al, Ni, Ti, and Ni-based alloys can be deposited by cold spray (Papyrin 2001); even cermets (Karthikeyan et al. 2001) and ceramics (Karthikeyan et al. 2002) can be embedded into a substrate to form a thin layer coating in cold spray.

The most important parameter in the cold spray process is the particle velocity prior to impact on the substrate (Alkimov et al. 1990). For a given material, there exists a critical particle velocity that must be achieved. Only particles whose velocities exceed this value can be effectively deposited, in turn producing the desired coating. Conversely, particles that have not reached this threshold velocity contribute to the erosion of the substrate (Alkimov et al. 1990). Naturally, critical particle velocities vary according to the particular

Introduction to Low Pressure Gas Dynamic Spray Roman Gr. Maev and Volf Leshchynsky

ISBN: 978-3-527-40659-3

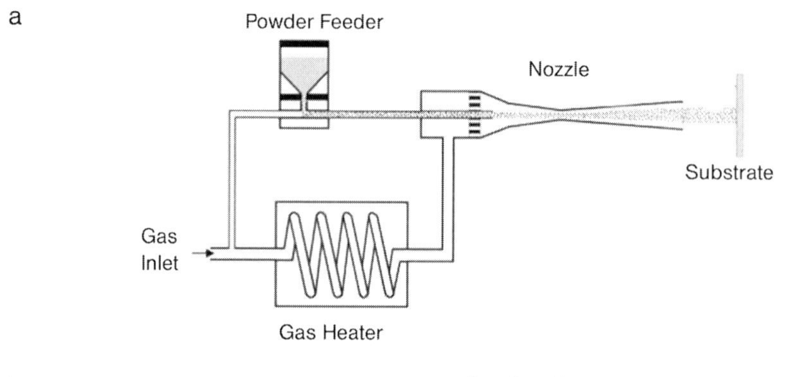

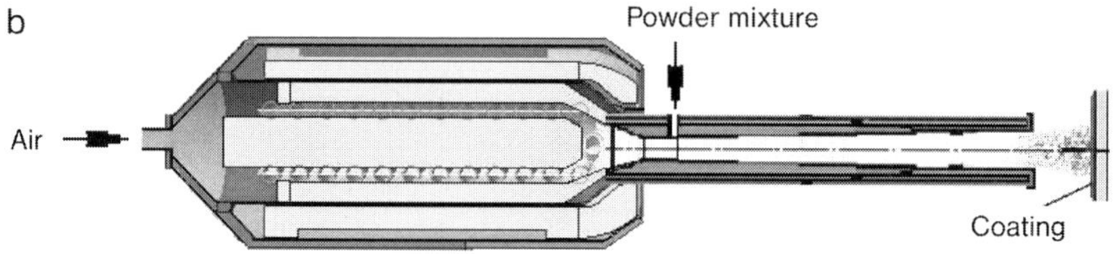

Fig. 1.1 Gas Dynamic Spray operating systems
Micron-sized particles are accelerated to high velocities by entraining the particles in the flow of a supersonic nozzle: (a) axial injection (Assadi et al. (2003)), (b) radial injection.

spray material that is chosen. In fact, critical velocities are dependent on both the size of the particles and their size distribution of particles (Alkimov et al. 1990, Van Steenkiste et al. 1999) and the particular substrate material (Gilmore et al. 1999). For example, it has been reported that the critical particle velocities of Cu, Fe, Ni, and Al are approximately 560–580, 620–640, 620–640, and 680–700 m/s, respectively (Alkimov et al. 1990).

The particle velocity, on the other hand, is determined by the nature, operating pressure, and temperature of accelerating gas, as well as by nozzle design (Gilmore et al. 1999, Dykhuizen et al. 1998). Material parameters such as density, particle size, and morphology influence the acceleration and subsequent deposition behavior of the particles (Van Steenkiste et al. 1999, Gilmore et al. 1999). It has been demonstrated that a dense coating can indeed be deposited through cold spraying if these spray conditions contribute to particle velocities that exceed the critical particle velocity (McCune et al. 1996). However, even in the case of deposition efficiencies greater than 80%, Karthikeyan et al. (2000) reported that porous titanium coatings may be formed.

A unique feature of cold gas-dynamic spraying is its ability to generate a wide range of deposition layer thicknesses ranging from tens of microns up to a centimeter (Li et al. 2004). In this regard, this process extends beyond the concept of "coating" a substrate, providing a means for developing three-dimensional structures (McCune et al. 1996). In fact, in a recent paper by

Gabel and Tapphorn (1997) a material forming process is discussed which utilizes a low-temperature gas stream to accelerate solid particles into a jet. Further, Bhagat et al. (1997) reported the use of cold spray processes for the production of nickel–bronze layers that may be employed in wear resistance applications.

The acceleration of powder particles in gas dynamic spraying is realized within the GDS nozzle. Here compressed gases (having inlet pressures up to 30 bar) flow through a converging/diverging nozzle, thereby developing supersonic velocities. The powder particles are metered into the gas flow immediately upstream of the converging section of the nozzle where they are accelerated by the rapidly expanding gas. It is perhaps obvious that compressed gases that are preheated are able to achieve higher gas flow velocities in the nozzle. However, even with preheat temperatures as high as 900 K, as used by Dykhuizen et al. (1998), the gas rapidly cools as it expands in the diverging section of the nozzle. Hence, the duration of particle interaction with the hot gas is relatively brief, and the temperatures of the solid particles remain substantially below that of the gas (Dykhuizen et al. 1998). Thus, the primary function of the carrier gas is to only to accelerate the particles to above the critical velocity.

The actual mechanism by which the solid-state particles deform and bond has yet been fully characterized. It seems possible that plastic deformation may disrupt thin surface films, such as oxides, and provide intimate conformal contact under high local pressure, thus permitting bonding to occur (Dykhuizen et al. 1999). Though unproven, this hypothesis is consistent with the fact that a wide range of ductile materials, such as metals and polymers, have been cold-spray deposited. Experiments with materials, such as ceramics, on the other hand, have proven unsuccessful unless sprayed along with a ductile powder material. If indeed this is the case, the minimum critical velocity corresponds to the kinetic energy that is necessary to bring about powder material plastic flow during deposition. Calculations by Schmidt et al. (2003) indicate that typical kinetic energies found at impact are less than that which is required to melt the particles. Micrographs of cold-sprayed materials (Dykhuizen et al. 1998) also reveal that the deposition mechanism is primarily, if not entirely, a solid-state process.

Because sufficient particle velocity is essential for successful cold-spray deposition, it is critical to fully understand the relative influence of process variables (such as gas inlet pressure, temperature, and nozzle geometry) on particle velocity. For this reason, discussions of the basic governing equations, computational results, and comparisons with experimental datum, will be take place primarily with respect to the relationship between these variables. It is hoped that this form of analysis will bring about a deeper understanding of the GDS process, and that this book may ultimately serve to widen the study and application of gas dynamic spraying.

1.2
Overview of Competitive Technologies

1.2.1
Coating Characterization

Today numerous surface treatment procedures are used to modify the surface properties of various materials without altering their bulk characteristics. These techniques are used within industrial environments to improve resistance to corrosion, wear, fatigue, and heat. In order to properly analyze and, in turn, distinguish between these methods, it is first beneficial to differentiate between thin and thick layers.

Thin coatings have more been developed on a considerably wider scale. These include techniques such as CVD or PVD (chemical or physical vapor deposition), plasma spraying, ion-assisted deposition, magnetron-sputtered deposition, and chemical deposition under a laser beam. Their thicknesses, being generally less than a few micrometers, have no direct effect on the mechanical characteristics of the treated material; they do not alter such aspects as yield strength or ultimate tensile strength.

On the other hand, thick coatings, including cladding and welding, have thicknesses varying from a few hundred micrometers to several centimeters. They generally serve not only as protective layers, but also add mechanical strength. Most of the problems encountered when welding or cladding materials concern the heat-affected zone (HAZ) and result from the heat generated by a laser (Kalla 1996) or electron beam (Yilbas et al. 1998), or alternatively as a result of capacitive discharge (Simmons and Wilson 1996). In processes characterized by high energy densities such as plasma arc, electron beam welding, and laser beam welding (Sun and Karppi 1996, Tjong et al. 1995), the HAZ is considerably smaller than that in more conventional processes such as gas metal arc and submerged arc welding (Cullison 2001, Kannatey-Asibu 1997). This is primarily due to very short interaction times. Even with a small HAZ, incompatibilities may still exist due to differences in the physical properties of the materials such as fusion points and thermal expansion coefficients (Grodzinski et al. 1996, Conzone et al. 1997). This may ultimately induce cracking (Conzone et al. 1997, Wang et al. 2000), high residual stresses (Wang et al. 2000, Conzone 1997), or brittle intermetallic compounds (Wang et al. 2000).

Thermally sprayed coatings are used extensively within a wide range of industrial applications. These techniques generally involve spraying molten powder or wire feedstock that has been melted using oxy-fuel combustion or an electric arc (plasma). The molten particles are subsequently accelerated toward a properly prepared (often metallic) substrate. Solidification occurs at rates akin to those obtained in rapid solidification technology, resulting in deposits that have ultrafine grain structure, with similar microstructure characteristics.

Thermal spray is a highly flexible process which is able to generate dense, tenaciously bonded coatings with low porosity. It is commonly used to shield surfaces from wear, high temperatures, and chemical attack, along with environmental corrosion protection in infrastructure maintenance engineering. Typical materials that are sprayed in this method include most metal alloys and ceramics. In fact, virtually any material can be thermal sprayed if it does not decompose prior to melting.

Various thermal spraying methods, such as flame spraying (including high-velocity air-fuel (HVAF) and high-velocity oxy-fuel (HVOF) thermal spray devices), plasma spraying, anelectric arc sprayingd , have been used to coat both metallic and nonmetallic surfaces. A general overview of existing spraying techniques is given below.

1.2.2 Flame Spraying

Although *flame spraying* was one of the earliest thermal spray processes to be developed, its usefulness endures even today. In this method, metals, ceramics, or cement-like materials are deposited onto a substrate by means of the combustion reaction. The flame spray device includes a combustion chamber that receives a mixture of fuel (e.g., propylene or propane) and oxidant (e.g., oxygen or air) in the form of a high-temperature, high-pressure gas stream. This combustion stream is initially directed into a flow nozzle where the spray material (e.g., a powder, a solid rod, or a wire) is introduced and at least partially melted and "atomized" (Neiser et al. 1995). It is then propelled in this form to the target substrate. Depending on the design, particle streams may be accelerated up to supersonic or hypersonic velocities. This supersonic particle stream may be generated by a single- or two-stage combustion devices, or alternatively by those that produce steady-state, continuous detonations.

In HVOF techniques, propylene, propane, or hydrogen (depending on user requirements) is combined with oxygen in a proprietary siphon system in the front portion of the diamond jet gun. The thoroughly mixed gases are then ejected from a nozzle and ignited externally from the gun. Meanwhile, the coating material (in powdered form) is fed axially through the gun using nitrogen as a carrier gas. Thus, the ignited gases form a circular flame configuration which surrounds and uniformly heats the powdered spray material as it exits the gun and is propelled to the specimen surface.

It should be noted that due to the high kinetic energy transferred to the particles by the HVOF process, the coating material generally does not need to be fully melted. Instead, the powder particles are heated to a molten state and flatten plastically as they impact the surface. For this reason coatings are formed having more predictable chemistries that are very homogeneous and have a fine granular structure (Li et al. 1996).

Through this use of an efficient and controlled thermal output with a high kinetic energy, HVOF processes produce dense, very low porosity coatings that exhibit very high bond strengths (exceeding 12,000 PSI or 83 MPa), have low oxidization and yield extremely fine finishes, even without further processing. The resultant coatings exhibit low residual internal stresses and can be sprayed to thicknesses that are not normally associated with dense, thermal sprayed coatings.

1.2.3 Arc Wire Spraying

Arc wire spraying uses two metallic wires as the coating feedstock. These wires are electrically charged with opposing polarity and fed into the arc gun at matched, controlled speeds. When the wires are brought together at the nozzle, their opposing charges create enough heat to continuously melt their tips. Compressed air is used to atomize the subsequently molten material and, in turn, accelerates it toward the surface, forming the coating. In arc wire spraying, the weight of the coating that can be deposited per unit of time is a function of the electrical power (amperage) of the system, as well as of both the density and melting point of the wire. New arc wire spray coating systems have unique, advanced mechanisms such as synchronized "push–pull" wire feed motors which ensure consistent wire delivery.

1.2.4 Plasma Spraying

Plasma spraying is perhaps the most flexible of all of the thermal spray processes in that it can create sufficient energy to melt any material. Further, because it uses powder as the coating feedstock, the number of coating materials that can be used in the plasma spray process is almost unlimited. In this method, the plasma gun consists of a cathode (electrode) and an anode (nozzle) with a small gap forming a chamber between the two. As gases are passed through the chamber, DC power is applied to the cathode and, in turn, arcs across to the anode. The powerful arc is sufficient to strip the gases of their electrons, forming a state of matter known as plasma. As the coating material is introduced into this gas plume it is rapidly melted, forming a particle-plasma stream which is accelerated up to a hypersonic velocity and propelled toward the substrate.

In this process, various mixtures of nitrogen, argon, and helium (usually two of the four) are used in combination with the electrode current to control the amount of energy produced by a plasma system. Since the flow of each of the gases and the applied current can be accurately regulated, repeatable and predictable coatings can be obtained. In addition, the point and angle at which the material is injected into the plume, as well as the distance between the gun and the target, can be precisely controlled. This provides a high degree of flexibility in developing appropriate spray parameters for materials having a very wide range of melting temperatures.

1.2.5
Rapid Prototyping

In materials processing specific parts are manufactured to particular specifications from a variety of materials. For viable product realization in the marketplace, a spectrum of issues regarding performance, cost, and environmental impact must constantly be addressed. Novel processing methodologies, as well as the innovation of new materials, provide opportunities for product development, improvements in product performance, reductions in cost, and minimization of environmental impact through the complete lifecycle (Zhang et al. 2001). One of these innovations, namely additive manufacturing, has allowed for product modification through controlled consolidation, layering, and/or coating, in turn leading to superior product attributes. In fact, the idea of fabricating components through the addition of material led to the establishment of the rapid prototyping industry approximately 15 years ago (Zhang et al. 2003b).

Rapid prototyping was initially considered a tool for the rapid development of 3D models and prototype parts, used almost exclusively to verify the design feasibility rather than for functional application. Currently available commercial rapid prototyping techniques, such as stereolithography, selective laser sintering (SLS), laminated object manufacturing, fused deposition manufacturing, and 3D printing are able only to produce prototypes using wax, plastic, nylon, paper, polycarbonate materials, etc. There are, however, many difficulties that must be addressed when attempting to replicate these processes techniques using materials with high melting temperature such as metal (Li et al. 2000). These include poor material strength, high porosity, oxidation, warping, “step” effects, etc. There is, however, great interest in further developing current materials additive manufacturing (MAM) techniques or alternatively, to create new methods which are capable of directly producing metal parts.

The MAM techniques that can be used to directly manufacture metal parts can be divided into two groups: 3D cladding (Mazumder et al. 1997, Milewski et al. 1998) and 3D welding (Schmidt et al. 1990). In 3D cladding, a laser beam creates a weld pool into which powder is injected and melted (Choi et al. 2001). The substrate is then scanned by a laser-powder system in order to trace a cross-section which, upon solidification, forms the cross-section of a part. Consecutive layers are then additively deposited, thereby producing a 3D component (Jeng et al. 2000). An example of this technique, known as laser engineering net shaping (LENS), was developed by Sandia National Laboratory. In this method metal components are fabricated directly from CAD solid models, thus further reducing the lead times for metal part fabrication. A similar process by the name of *directed light fabrication* (DLF) is under development at the Los Alamos National Laboratory (Milewski et al. 1998).

1.2.6 Plasma Deposition Manufacturing

Plasma deposition manufacturing (PDM) is a newly developed direct metal fabrication process that belongs in the category of 3D welding. In this technique, a supply head delivers a well-defined flow rate of metal powder, which is deposited in a molten pool formed by controlled plasma heating. In this way, full strength parts with the single or multifarious materials can be built up, layer by layer, by melting and rapid solidifying the feed material into a particular shape.

The PDM process has many advantages over traditional material subtractive technologies: (1) conventional metal melting and casting techniques frequently require considerable time and expenditure with respect to mold production, particularly in low volume manufacturing. PDM, by contrast, allows for the direct production of individual parts in a single step. Also, the manufacturing materials may be varied either between parts or within a single component. (2) Most MAM techniques are limited to the production of parts composed of plastic, wax, or paper; thus, the required engineering properties cannot always be obtained. The PDM process, on the other hand, allows for the direct fabrication of metal or end-use-material components. Further, the deposition rate and particular materials used in PDM can be adjusted through a wide range, making this technology extremely flexible. (3) PDM can be used as an effective remanufacturing tool, helping to increase product lifecycle.

There is, however, a primary disadvantage to the PDM process in that the effects of local melting due to the high temperatures cannot be avoided. For this reason it seems reasonable (and quite advantageous) to apply the principles of PDM to GDS. In this case, the solid-state bonding mechanism will allow the deposited material to retain its original structure.

The most efficient among the thermal spraying methods is plasma deposition, combining a high-temperature gas jet with the high productivity of a coating process. Unfortunately, while plasma spraying can produce high-quality coatings, this method suffers from several minor limitations. Firstly, the necessary apparatus is relatively complex and expensive. Secondly, although plasma spray processing is a well-established method for forming thick coatings, in coatings thinner than 100 μm it is difficult to control material properties. Also, in spite of the plasma jet velocity being about 1000 to 2000 m/s, and the particles being accelerated up to 50 to 200 m/s (and even up to 350 m/s in extreme cases), the acceleration process is not yet efficient enough. The primary imperfection associated with plasma spraying, however, is its high processing temperature, making a number of substrate materials unfit for treatment due to local heating, oxidation, and thermal deformations. Further, the high-temperature jet that is incident on the product surface intensifies chemical and thermal processes, causing phase transformations and the appearance of oversaturated and nonstoichiometric structures which bring about structural changes in the substrate material. Also, the high cooling rates result in the hardening of these heated materials. And finally, the liberation of gases during crystallization brings about both porosity and the appearance of microcracks.

In short, thermal spraying methods have the following disadvantages: (1) A high level of thermal and dynamic effects on the surface being coated; (2) substantial changes in the properties of the material to which the coating is being applied (i.e., electrical and heat conductance, etc.); (3) changes in the material structure result from the chemical and thermal effects of the plasma jet and the hardening of overheated melts; (4) ineffective powder particle accelerations due to the low density of the plasma; (5) intensive evaporation of fine powder fractions having sizes ranging from 1 to 20 μm; (6) insufficient cost-per-mass ratio of the applied material; (7) reduced equipment portability due to the high-temperature requirements of the method.

1.2.7 Explosive Cladding

The processes which are best able to avoid the problems associated with thermal changes appear to be spot welding (Turgutlu et al. 1995, 1997), impact welding (Date et al. 1999), or explosive cladding (EC). These methods allow for flyer plate velocities that exceed plasma and detonation gun spraying – both of which result in poor bonding and increased surface roughness (Li Ohmori 1996). Explosive bonding is considered to be a solid-state welding process since melting is not required at the interface of the two materials in order for bonding to occur (Zimmerly et al. 1994). Moreover, as there is little to no diffusion between the two metals or alloys, each retains its intrinsic properties (El-Sobky 1993). This process makes it possible to join materials having greatly differing melting points without the need for intermetallic compounds – even those having the historic reputation of being incompatible (e.g., titanium and stainless steel, nickel and aluminum, etc.) (Gerland et al. 2000). This is contrary to what occurs in other techniques such as laser cladding (Kalla 1996, Yilbas et al. 1998).

Explosive cladding (EC) is a solid-state fusion welding process in which the joining of two metals is accomplished through the application of a shock pressure that results from the detonation of an explosive pack (Raybould et al. 1981). Material interaction at the mating surfaces during EC produces a "surface jetting" effect, ultimately resulting in a strong bond being formed between the metals (Raghukandan 2003). As it turns out, an effect analogous to this "surface jetting" is the main feature of GDS processes, suggesting that the EC and GDS are of the same fundamental nature. One should further note that the impact velocities of EC vary in the range of 1400 to 3900 m/s, while the particle velocities of GDS are known to have lesser values ranging from 500 to 1600 m/s for different carrier gases. Therefore, the intensity of "surface jetting" or jet flow in the GDS process seems to be less than that of EC. Unlike in EC, however, the particle impact velocity in GDS may be precisely controlled; it is through this control that the structure and material properties of the deposited layers may be managed.

1.3 Concluding Remarks

A comparison of competitive spraying technologies clearly shows that cold gas dynamic spraying has been established as a viable coating technology in the thermal spray processes family. The dense and oxide-free coatings that may be produced through GDS have brought about a multitude of new applications which, up to now, have not been feasible using traditional processes. As a result of many years of targeted and process-oriented groundwork by various researchers and manufacturers alike, both high- and low-pressure GDS engineering have reached advanced technological levels, facilitating efficient operation with a high degree of process reliability. As a result, practical solutions may be readily observed within such industries as automotive and electronics manufacturing. Despite this great progress, however, the full potential of high-pressure and low-pressure GDS has not yet been fully realized.

2 Impact Features of Gas Dynamic Spray Technology

2.1 Impact Phenomena in GDS

2.1.1 Main Features

It is perhaps obvious that particle impact is the main feature which governs the consolidation process in cold spraying. In fact, particle impact phenomena have been studied for many years, for a wide range of particle velocities (Klinkov et al. 2005). The nature of layer formation in the cold spray process, on the other hand, is far from being fully understood. Currently, extensive theories explaining the processes of cold spray deposition are being developed (Karthikeyan 2005). At this point in time, however, the actual mechanism by which the powder particles are consolidated, deformed, and bonded with the substrate is not well understood, particularly with respect to GDS processes having small particle velocities or low pressures. For this reason it is yet necessary to closely scrutinize the impact interactions of the particles in hopes of demonstrating how these main features influence the consolidation process.

The two main phenomena that may be used to characterize the impact are: (i) the interaction of a single particle with the surface of the substrate, or a previously bonded particle, and (ii) the shock compression of the powder layer (Dykhuizen et al. 1999). While the individual particle approach is being intensively examined and modeled, the shock compression model requires further scrutiny. The shock-compression processes that are realized by GDS seem to produce highly activated powder mixtures. This occurs through defect generation and grain size reduction via fracturing and/or the formation of subgrain structures during interparticle sliding, severe particle deformation, and consolidation (see Fig. 2.1). These effects result in significantly increased mass transport rates and enhanced chemical reactivity in the powders, creating new paths for the movement of point defects at the interface. In particular,

Introduction to Low Pressure Gas Dynamic Spray Roman Gr. Maev and Volf Leshchynsky

ISBN: 978-3-527-40659-3

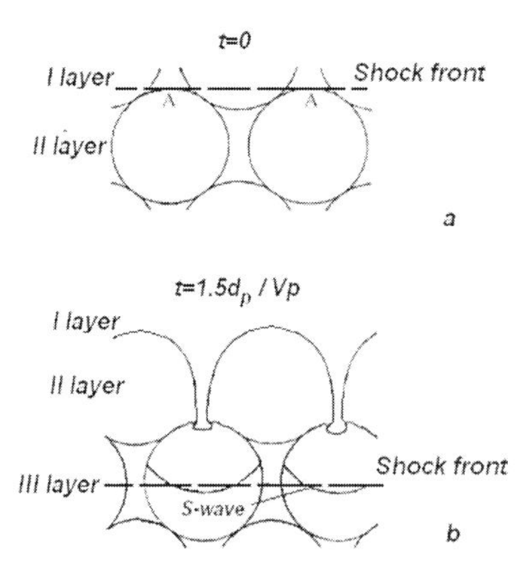

Fig. 2.1 Shock compaction of a spherical particles layer. The particles of diameter d_p are subjected to the impact by a powderladen jet with velocity V_p, which coincides with the passage of the shock-wave front, causing distortion of the particle shape. $a - t = 0$; $b - t = 1.5d_p/V_p$.

the interaction of the powder mixture components can be accelerated by the presence of dislocations, shorter diffusion distances, more intimate cleansed surface contacts, and higher packing densities, all of which favor solid-state diffusion reactions.

Microparticle impact with surfaces continues to be studied because of its involvement in many important processes such as surface contamination, powder transport, coating application, etc. Although a considerable number of studies have been conducted within the last 30 years, an overall understanding of the deposition process has yet to be realized. In fact, although the fate of microparticles after impact with a surface can be satisfactorily explained by several models, none of them have been entirely validated by experimental data. This is because these models require information that is not easy to quantify, such as the amount of elastic and plastic deformation, the effective inertial mass of the substrate, the bulk and surface properties of the material, etc. (Li et al. 1999). Ideally, these data should be based not only on the properties of the bulk material, but also on those of the microparticles. At present time, however, this information is difficult to obtain. For this reason it is perhaps useful to develop models which instead establish the general relationships between these variables.

2.1.2
Rebound and Erosion Processes

In principle, particles impinging onto the specimen can rebound from, stick to, or penetrate into a bulk substrate. Oftentimes, the impact of a particle on a surface leads to the deformation or destruction of both the particle and the

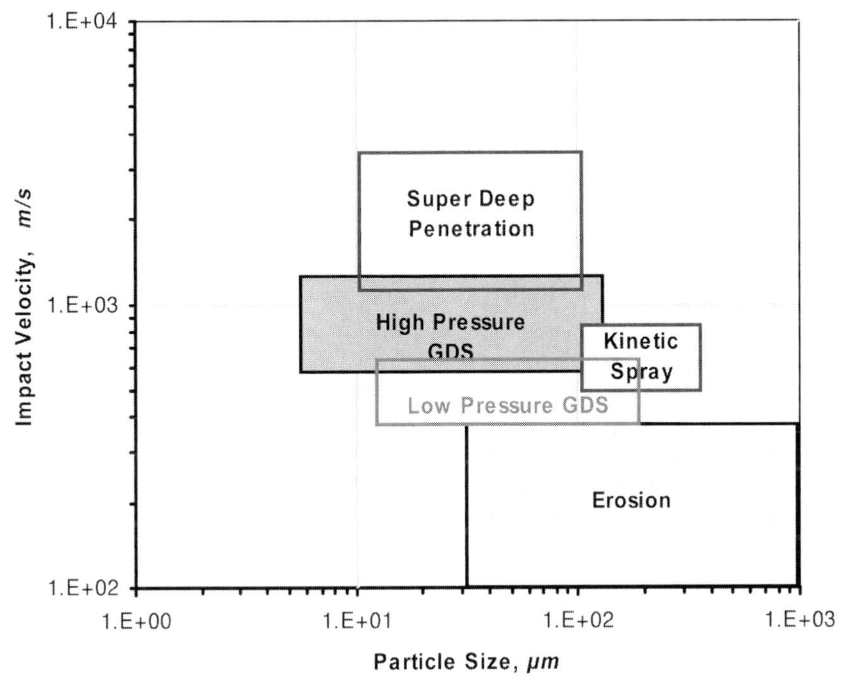

Fig. 2.2 The relative impact velocity–particle size of high pressure, kinetic spray, and low-pressure GDS.

solid body. Klinkov et al. (2005) have classified the results of particle impact on plane surfaces. This classification is based on two important dimensional parameters which may be used to characterize particle impact: the impact velocity, v_p, and the diameter, d_p, of impinging particles (Fig. 2.2). These impact parameters for certain GDS processes may be identified through observation of the results published in many papers. From these works one can readily see that the regions of the each GDS process (high pressure, low pressure, or kinetic GDS) strongly depend on both v_p – and, in turn, on the kinetic energy of the particle – and d_p. Further, it should be noted that all of the GDS processes occur with particle velocities ranging from 10^2 to 10^3 m/s.

The indentation and erosion of metallic plates due to the impact of particles has been observed for a wide range of particle sizes and velocities (Zukas 1990). Maximum erosion for a ductile target, however, has been shown to occur at incidence angles of approximately 70–80° to the normal. It is generally believed that this indentation and erosion takes place as a result of a combination of plowing and cutting actions. In this regime, plastic flow and friction are the key mechanisms of deformation and energy dissipation. To gain insight into the fundamental mechanisms underlying solid-particle erosion, Molinari and Ortiz (2002) performed detailed finite-element simulations modeling the impact of spherical particles on metallic plates. It was found that the primary process governing this erosion is adiabatic shear band formation due to plastic deformation and friction, heat generation, and thermal softening.

Upon examining the dependence of rebound velocity on impact velocity (Fig. 2.3), three particle velocity zones are clearly visible (Molinari and Ortiz 2002). The nondimensional velocity parameter, v_r/v_p, and the average particle temperature increase function, ΔT_r (in K), on the other hand, may only be divided into two regimes. In fact, a logarithmic approximation of the ratio v_r/v_p gives $v_r/v_p = -0.21\ \ln(v_p) + 1.8$ for the low particle velocities (up to 600 m/s), and $v_r/v_p = -0.29\ \ln(v_p) + 2.3$ for higher particle velocities (above 600 m/s). The large difference between the coefficients of these equations suggests that there are two distinct mechanisms of energy dissipation that take place for particles in these velocity ranges. This behavior can also be detected when measuring the average particle temperature increase, ΔT_r. Its specific dependence on particle velocity can be seen in Fig. 2.3, where these same low and high impact velocity groups can be observed. The logarithmic approximations for the temperatures in these velocity groups are $\Delta T_r = 67.2\ \ln(v_p) - 361.4$ K for low impact velocities, and $\Delta T_r = 376.6\ \ln(v_p) - 2391.9$ K for high impact velocities. These trends appear to be related to a change in the particle impact mechanism from nearly elastic to plastic; this change occurs at the critical velocity of 600 m/s. A simulation of the change in contact forces by Li et al. (1999) is shown in Fig. 2.4 for low particle velocities (normal velocity of 5.2 m/s for the molybdenum surface). From this simulation, it is abundantly clear that the Hertzian dissipation force is relatively small, implying nearly perfect elastic impact. It is further observed that the adhesion dissipation force is prevalent during the rebound phase, and thus is primarily responsible for the total energy lost during the impact.

Due to plastic deformation, energy losses result in a decrease in the velocities of rebound particles. In this regime, the deformation occurs too rapidly for effective heat conduction, resulting in thermal softening and adiabatic shearing, and ultimately reducing the bearing capacity of the material. The sharp nature of this transition is suggestive of a critical phenomenon and is typical of shear localization instabilities (Molinari and Ortiz 2002). The region of super deep penetration (SDP) phenomena coincides with particle velocities ranging from 1.2×10^3 to 2.5×10^3 m/s. The main characteristic of SDP is the penetration of a small fraction of particles ($\sim$0.1%) to a specified depth, 10 d_p, within the substrate. This occurs for particle velocities exceeding 10^3 m/s, provided that the hardness of the particles is greater than that of the substrate (Kiselev 2002). Finally, an area characterized by hypervelocity impact ($v_p > 2 \times 10^3$ m/s) lies beyond consideration, see Fig. 2.2.

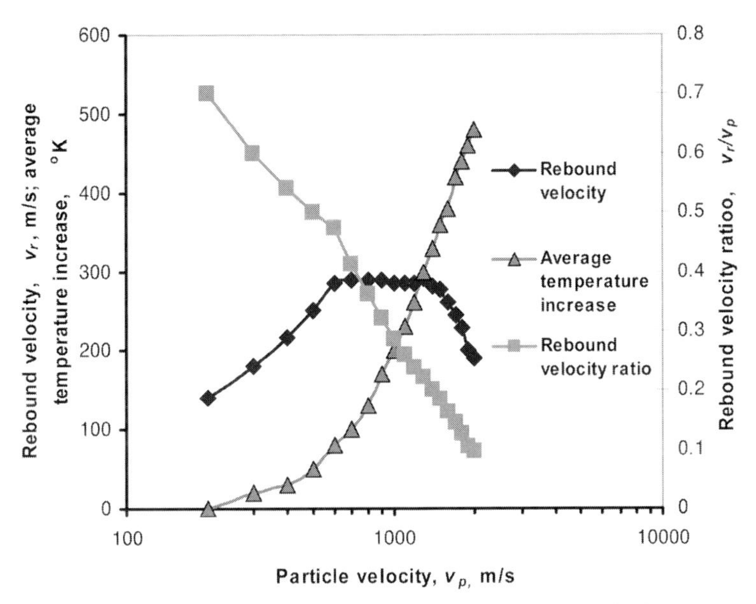

Fig. 2.3 Normal incidence impact: rebound velocity, rebound velocity ratio, and an average temperature increase vs. particle impact velocity, after Molinari and Ortiz (2002). Spherical particles are 9.5 mm in diameter $\sigma_f = 1.43$ GPa. Impingement with a mild steel plate.

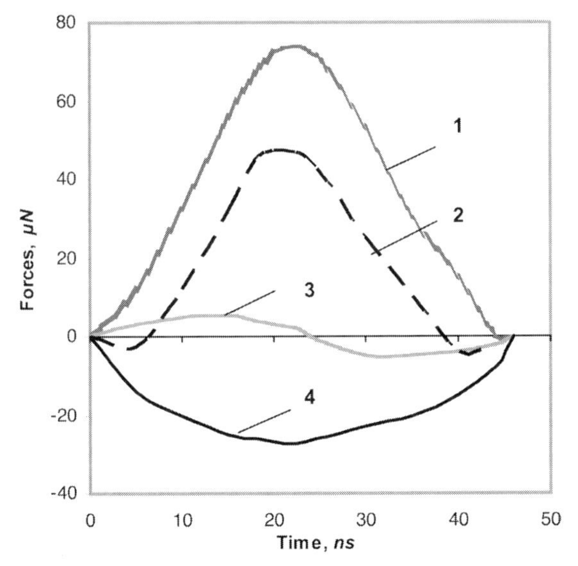

Fig. 2.4 Contact force variation with contact time, after Li et al. (1999). (1) Hertzian force, (2) total force, (3) adhesion dissipation force, and (4) adhesion force.

2.1.3 GDS Processes

In the region where GDS processes are realized, particles impact and bond with the surface for specifically defined particle velocities. Many authors (Papyrin 2001, Alkhimov et al. 2001, Kreye and Stoltenhoff 2000) consider the cold spray phenomenon to be strictly governed by the interaction of particles with the substrate. However, there is a distinct lack of literature describing and evaluating collective particle behavior in GDS. As discussed in the previous section, the interaction of incoming particles with the substrate can result in particles that rebound from (rather than adhere to) the surface. In order to fully understand the GDS process, then, it is necessary to characterize the effect of preliminary impacts that do not result in material adhesion. Klinkov et al. (2005) suggests that these impacts lead to the activation of the surface. Succeeding particles impinging on an activated surface area are then more likely to bond to the substrate. At present time, however, proper characterization of this surface activation with respect to, for example, the nature and quantity of preliminary impacts is difficult at best. For this reason, its physical mechanism is not fully understood.

As described earlier, in GDS high particle velocities are obtained through accelerating an expanding gas stream to velocities in the range of 500–1200 m/s by means of a converging diverging de Laval type nozzle. In this process, the gas is heated only to increase the particle velocity and to increase particles deformation upon impact. Thus, powder layers form as a result of the kinetic energy of the particles as they hit the substrate; the bonding of particles in GDS forms as a result of extensive plastic deformation and related phenomena at the interface. In general, for the particle to be bonded, its kinetic energy must be transformed into heat and strain energy within both the particle and substrate. For this reason inelastic collision processes which involve the plastic deformation of the particle and substrate are the main phenomenon causing particle consolidation. Such plastic deformation can be roughly approximated by the flattening of spherical incident particles into a plate-like structure having aspect ratios between 3:1 and 5:1 (Van Steenkiste et al. 2002). In order for sufficient plastic deformation to take place, the contact stress at the interface must exceed the yield stress of the particle. Thus, the powders that are typically employed in most GDS processes consist of metals with relatively high ductility. Hence, the fundamental theories of plasticity provide the scientific basis for particle impact analysis (Segal 2002). Moreover, the localization of deformation plays a very important role in this phenomenon, and will be appropriately pursued in the following chapters.

The bonding mechanisms in GDS can be compared to those in processes such as explosive welding and powder compaction. In explosive cladding, an area of the interface undergoes a severe deformation which is characterized by adiabatic shear bands, highly elongated grains, recrystallized grains, ruptured oxide films, and even resolidified microstructures (Hammerschmidt and Kreye 1981). In explosive powder compaction, dense powder materials can be produced through specifically chosen combinations of shock, pressure, and duration (Grujicic et al. 2004). As with explosive cladding, successful bonding in powder compaction is dependent on the critical conditions for extensive plastic deformation at either the particle/particle or particle/substrate interfaces (Nesterenko 1995). However, Assadi et al. (2003) point out that despite the similarities that exist among these processes, it has not yet become clear as to what extent the theories describing explosive cladding or shock wave powder compaction can be applied to GDS.

In light of the above review, then, these two approaches must then be considered together, with their principles coupled together to achieve a deeper understanding the GDS process. In particular, to develop an adequate theory describing low pressure GDS, one must consider (i) one particle impact, and (ii) powder ensemble behavior during impact.

2.2 One Particle Impact in GDS

2.2.1 Shear Localization Phenomenon

In a fundamental study by Assadi et al. (2003), comprehensive experimental and computational finite element analysis of a copper cold-spraying deposition process was undertaken. The results of this study suggest that the cold spray bonding mechanism can be attributed to adiabatic shear instability occurring at the particle/substrate or particle/deposited material interfaces for high impact particle velocities. In the papers of Gruijicic et al. (2003, 2004), the analysis of Assadi et al. (2003) is extended to several other metallic material systems. Further, a more detailed analysis of the susceptibility of metallic materials to adiabatic shear instability during cold spray is undertaken. To better understand the role adiabatic shear instability (or shear localization) plays within the bonding and consolidation processes of GDS, it is beneficial at this time to discuss the basic fundamentals of this phenomenon.

Adiabatic shear instability and the associated formation of shear bands have been considered in detail by Wright (1992, 2002), Bai (1982), Molinari (1997), and others. When metals are subjected to large deformations, shear localization is an important aspect governing material flow. The process of adiabatic shear banding was analyzed by Molinari (1997) from the experimental results

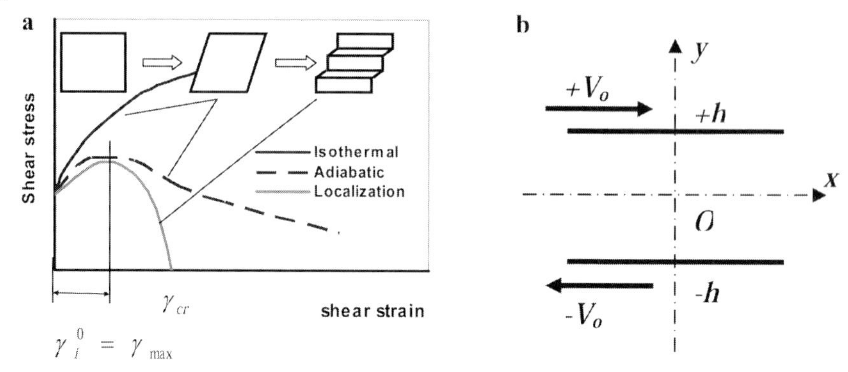

Fig. 2.5 Flow localization processes. (a) Schematic of the stress–strain curves, after Grujicic et al. (2003); (b) geometry used for shear analysis.

of Marchand and Duffy (1988) using thin-walled tubes that were twisted to high strain rates using a Kolsky bar configuration.

As shown in Fig. 2.5, there are three primary flow localization processes. In the first (a typical strain-hardening material under nonadiabatic conditions) the shear stress increases due to strain hardening (the curve labeled "Isothermal"). In the second, the plastic strain energy is dissipated as heat under adiabatic conditions, resulting in a temperature increase, in turn causing material softening whose effect overwhelms that of strain hardening. The primary change in the mechanical behavior of material is indicated by the occurrence a shear stress maximum in the flow curve (the curve labeled "Adiabatic"). It is during this stage that the development of flow heterogeneity may be observed, indicating the onset of a weak instability process (Molinari 1997). The third type of plastic flow is characterized by the marked development of strain localization, resulting in the formation of an adiabatic shear band that eventually propagates along the circumference of the specimen (Fig. 2.5). The fluctuations in stress, strain, temperature or microstructure, and the inherent instability of strain softening can give rise to plastic flow (shear) localization (Grujicic et al. 2004). Thus, shearing and heating (and subsequent softening) becomes highly localized, while the material strain and heating in surrounding regions remain negligible. This, in turn, causes the flow stress to rapidly decrease (the curve denoted "Localization").

In order to utilize the adiabatic shear instability modeling techniques, it is essential to understand the flow behavior of the material under processing conditions. Constitutive equations, which relate the flow stress σ in terms of strain ϵ, strain rate $\dot{\epsilon}$, and temperature T,

$$\sigma = f(\epsilon, \dot{\epsilon}, T) \tag{2.1}$$

are often used in deformation problems. To effectively evaluate the deformation and shear localization parameters, the flow stress σ data have to be defined in terms of strain ϵ, strain rate $\dot{\epsilon}$, and temperature T. In order to examine the various deformation models that exist for the regions of flow instabilities during deformation, the following dimensionless parameters must be determined from the measured flow stress data (Murty et al. 2000). The strain rate sensitivity parameter,

$$m = \frac{\partial \ln \sigma}{\partial \ln \dot{\epsilon}} \tag{2.2}$$

The flow softening rate,

$$w = \frac{1}{\sigma}\frac{\partial \sigma}{\partial \epsilon} = \frac{\partial \ln \sigma}{\partial \epsilon} \tag{2.3}$$

The flow localization parameter for plane strain compression,

$$\alpha = -\frac{\omega}{m} \tag{2.4}$$

The temperature sensitivity of flow stress $-a_s$,

$$a_s = \frac{1}{T}\frac{\partial \, ln \, \sigma}{\partial (1/T)} \tag{2.5}$$

In accordance with Chen and Batra (2000), we now consider the simple shearing deformations of an isotropic and homogeneous thermoviscoplastic body that is bounded by the planes $y = \pm h$ and sheared in the x-direction by prescribing velocities $v = \pm V_0$ on the upper and lower bounding surfaces (Fig. 2.5(b)). The bounding surfaces are assumed to be thermally insulated. The equations governing the deformations of the body are

$$\rho \ddot{\gamma} = \tau_{,yy}, \ \dot{\gamma} = v_{,y} \tag{2.6}$$

$$\rho c \dot{T} = kT_{,yy} + \beta \tau \dot{\gamma} \tag{2.7}$$

$$\tau = \tau_1 + \tau_2 \tag{2.8}$$

$$\tau_1 = B_1 e^{-(\beta_1 - \beta_2 ln \dot{\gamma})T} + B_2 \gamma^{1/2} e^{-(\alpha_1 - \alpha_2 ln \dot{\gamma})T} \tag{2.9}$$

$$\tau_2 = b \, \alpha_0 \, \mu_0 (1 - AT - BT^2) \lambda^{1/2} \tag{2.10}$$

$$\frac{\partial \lambda}{\partial \gamma} = f(\lambda, \dot{\gamma}, T) \tag{2.11}$$

Here ρ is the mass density, γ is the shear strain, τ is the shear stress, c is the specific heat, T is the absolute temperature, k is the thermal conductivity, and β is the Taylor–Quinney factor, representing the fraction of plastic work converted into heat. Also, a superimposed dot denotes the material time derivative and a comma followed by y represents the partial derivative with respect to y. Following the work of Zerilli and Armstrong (1997), and Klepaczko (1988), τ is written as the sum of τ_1 and τ_2. The portion of the shear stress given by Eq. (2.9) is a result of thermally activated dislocation interactions (Zerilli and Armstrong 1997). B_1, B_2, β_1, β_2, α_1, and α_2 are constants; B_1 and B_2 are related to the Gibbs free energy, the dislocation activation area at zero temperature, and the magnitude of the Burger vector. The portion of the shear stress given by Eq. (2.5) is dependent on the thermomechanical history of the plastic deformation through the internal variable λ, whose evolution is described by Eq. (2.6). This internal variable may be identified as the total dislocation density. In Eq. (2.10) b, a_0, μ_0, A, and B are constants.

One of the primary goals of an adiabatic shear instability analysis is to determine the critical strain and stress associated with material instability (Fig. 2.5(a)). The instability analysis undertaken by Chen and Batra (2000) showed that although the strain-rate hardening of the material does not directly influence material instability, it does have an indirect effect. For locally adiabatic deformations, material instability occurs only when thermal softening exceeds the combined effects of the material hardening due to plastic straining and the increase in dislocation density (Marchand and Duffy 1998). However, in the presence of heat conduction, higher values of the nominal strain rate delay the onset of the material instability. The modeling results presented by Chen and Batra (2000) evince the dependence of the instability onset strain γ_i^0, the corresponding shear stress τ_i^0, and the dislocation density λ_i^0 upon the nominal strain rate $\dot{\gamma}^0$ within a range of $\dot{\gamma}^0 = 10^2$–$10^5\ \mathrm{s}^{-1}$ (Fig. 2.6). These data allow one to clearly characterize the influence of the strain rate and temperature on both the position and value of the initial instability onset. For example, it can be seen that an increase in the strain rate shifts the adiabatic shear instability onset to lower strains and dislocation densities (Figs. 2.6(a) and (c)). Extrapolation of these functions up to $\dot{\gamma}^0 = 10^7\ \mathrm{s}^{-1}$ suggests an instability onset strain of $\gamma_i^0 = 0.13$; for $\dot{\gamma}^0 = 10^{-2}\ \mathrm{s}^{-1}$, $\gamma_i^0 = 0.21$. It is interesting to note that an increase in the initial temperature results in a stronger dependence of γ_i^0 upon $\dot{\gamma}^0$. At an initial temperature $T = 600$ K, the value of the instability onset strain $\gamma_i^0 = 0.063$ for $\dot{\gamma}^0 = 107\ \mathrm{s}^{-1}$ while $\gamma_i^0 = 0.3$ for $\dot{\gamma}^0 = 10^{-2}\ \mathrm{s}^{-1}$. Such a strong relationship between the initial temperature and the onset of material instability seems to indicate an intensive material softening at the beginning of the deformation process. This conclusion is confirmed by the fact that the shear stress is dependent on the strain rate at the onset of instability (Fig. 2.6(b)). In fact, the shear stresses remain constant at various temperatures for all strain rates (0.55–0.62 GPa). Also, it should be noted that the dislocation density at the onset of instability determines the level of material hardening, as well as the concentration of defects in the material. The dependence of the dislocation density on strain rate (Fig. 2.6c) can be characterized by $\gamma_i^0 = \gamma_i^0(\dot{\gamma}^0)$.

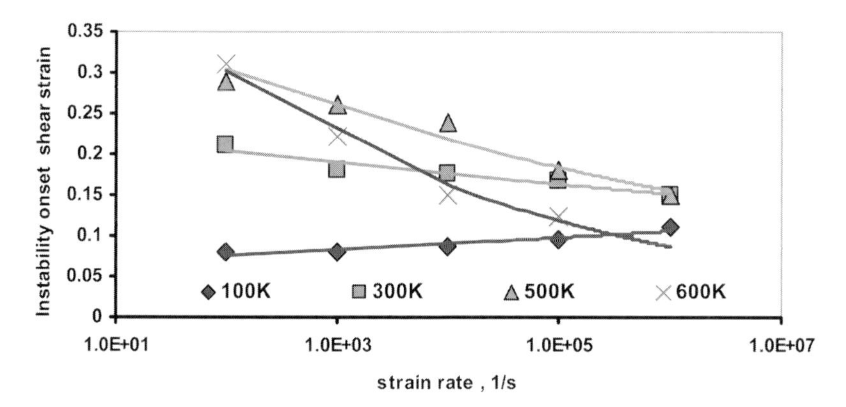

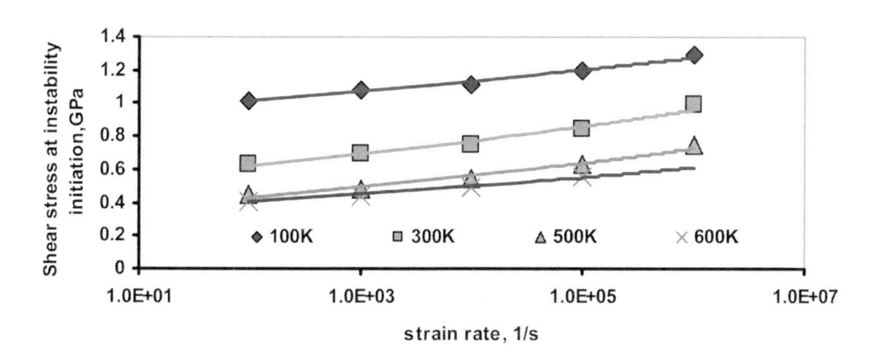

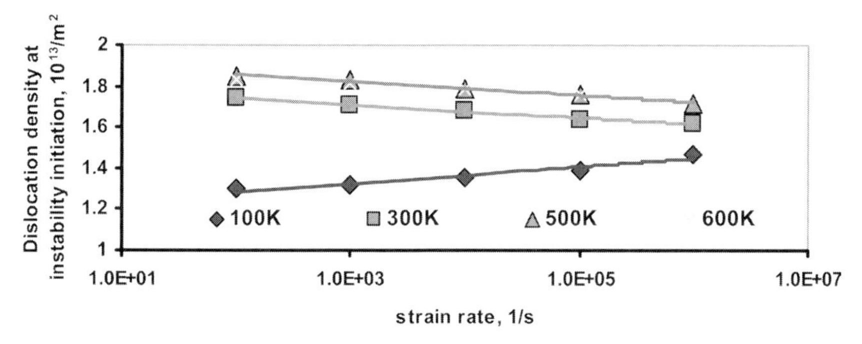

Fig. 2.6 Dependences of shear instability parameters, after Chen and Batra (2000). (a) Instability shear strain onset; (b) shear stress at instability initiation; (c) dislocation density at instability initiation.

Therefore, the primary factors which influence the development of shear localization processes are the strain rate and the initial temperature of material.

2.2.2
Adiabatic Shear Instability in GDS

The general equations described above are able to adequately describe shear instability within the GDS process with the particulars of this theory illustrated by Assadi et al. (2003) and Grujicic et al. (2004). In order to further this analysis, however, it is beneficial to discuss the model developed by Grujicic et al. (2004), and, in turn, to compare simulated results with experimental data.

To understand the onset of strain softening and adiabatic shear localization, a simple one-dimensional model is utilized by Grujicic et al. (2004). This technique, though relatively basic, is able to reveal the thermomechanical behavior of a small material element at the particle/substrate interface during collision. The model is presented in the terms of the equivalent plastic strain rate $\dot{\epsilon}^p$ (referred to as shearing by Schoenfeld and Wright (2003)), whose formal definition is given by

$$\gamma^p = \gamma_0^p + \int_0^t \dot{\gamma}^p dt$$

Here γ^p is the plastic shear strain and $\dot{\gamma}^p$ is an equivalent shear strain rate. Thus, $\dot{\gamma}^p = \dot{\epsilon}^p$.

This model of Grujicic et al. (2004) is based on the following governing equations: the equivalent plastic strain rate, $\dot{\epsilon}^p$, where

$$\dot{\epsilon}^p = \dot{\epsilon}_0^p \frac{\rho v_p^2}{2} \frac{1}{s} \tag{2.12}$$

with the boundary condition $\dot{\epsilon}^p(t=0) = \dot{\epsilon}_0^p \frac{(\rho v_p^0)^2}{2} \frac{1}{s_{\text{init}}}$.
The equivalent plastic strain, ϵ^p,

$$\epsilon^p = \int_0^t \dot{\epsilon}^p dt \tag{2.13}$$

The heating rate, $\dot{T}$,

$$\dot{T} = \frac{s\dot{\epsilon}^p}{\rho C_p} \tag{2.14}$$

with the boundary condition $\dot{T}(t=0) = \frac{s_{\text{init}}\dot{\epsilon}^p(t=0)}{\rho C_p}$.
The temperature, T,

$$T = \int_0^t \dot{T} dt \tag{2.15}$$

with the boundary condition $T(t=0) = T_{\text{init}}$.
And the particle velocity,

$$v_p = v_p^0 \left(1 - \frac{t}{t_c}\right) \qquad v_p(t=0) = v_p^0 \tag{2.16}$$

Here $\dot{\epsilon}_0^p$ is the strain rate proportionally constant, ρ is the (constant) material mass density, C_p is the (constant) specific heat, and a raised dot denotes the time derivative of a particular quantity.

A fundamental assumption is made in Eq. (2.12), where the equivalent plastic strain rate is taken to be inversely proportional to the equivalent plastic flow strength, s, and, in turn, a function of strain, strain rate, and temperature. This quantity is described by the well-known Johnson–Cook model (Johnson and Cook 1983):

$$s = [A + B(\epsilon^p)^n] \times \left[1 + m \ln\left(\frac{\dot{\epsilon}^p}{\dot{\epsilon}_0^p}\right)\right] \times \left[1 - \left(\frac{T - T_{\text{init}}}{T_{\text{melt}} - T_{\text{init}}}\right)^{\vartheta}\right] \tag{2.17}$$

where A, B, m, n, and ϑ are constants.

The model allows for effective evaluation of the plastic strain rate, heating rate, temperature, and equivalent stress in representative elements at the particle–substrate interface during particle impact. Similar calculations have been made by Assadi et al. (2003). A comparison of the modeling results is shown in Figs. 2.7 and 2.8 for equivalent normal stresses and strains. Here the dissimilarities in material deformation can be observed as follows: (i) the peak in the stress is more evident, and (ii) the transition to the adiabatic shear band formation process (corresponding to a sharp decline in the equivalent normal stress) is more clearly defined within the model developed by Grujicic et al. (2004). As shown in Fig. 2.9, the equivalent normal stress falls zero. This effect seems to be due to adiabatic softening in the area of localization; the specific duration of this process must be sufficient so as to adequately increase the temperature in this region. Calculated temperature variations with respect to impact times for various impact velocities are presented by Assadi et al. (2003). These are similar to the strain variations that are shown in Fig. 2.8 and are due to the strong interdependence of strains and temperatures in Eq. (2.17).

It should be noted that the assumptions utilized to determine the contact pressures in the models of Assadi et al. (2003) and Grujicic et al. (2004), though different, do not result in different values. In Grujicic et al. (2004), the contact pressure at the particle/substrate interface is assumed to be proportional to the kinetic energy of the particle per unit volume $\rho v_p^2/2$. This assumption, along with being quite reasonable, is readily utilized in most calculations. There is, however, a fundamental difficulty in determining ϵ_0^p due to the boundary condition

$$\dot{\epsilon}^p(t = 0) = \dot{\epsilon}_0^p \, \frac{(\rho v_p^0)^2}{2} \, \frac{1}{s_{\text{init}}}$$

In fact, because of the uncertainty of $s_{\text{init}} = f(v_p^0)$ it does not unambiguously define the value of $\dot{\epsilon}_0^p$. Thus, the primary requirement for the effective applica-

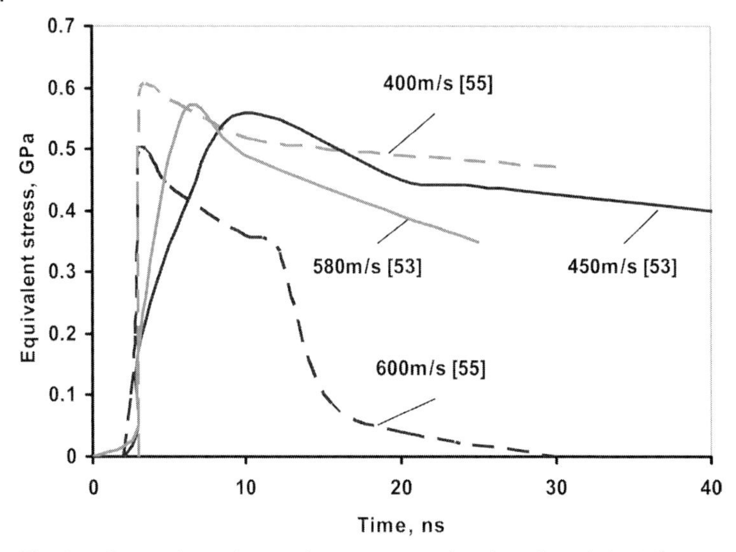

Fig. 2.7 Comparison of equivalent stresses: after Assadi et al. (2003); after Grujicic et al. (2004).

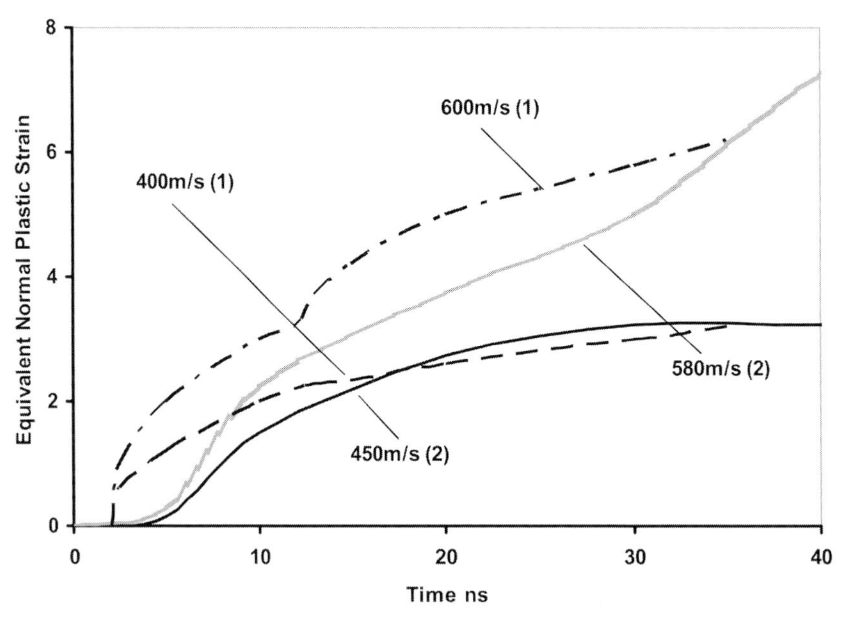

Fig. 2.8 Comparison of equivalent normal strains: (1) after Grujicic et al. (2004); (2) after Assadi et al. (2003).

tion of either model is the correct determination of the constants and variables that appear in Eqs. (2.12), (2.14), and (2.17).

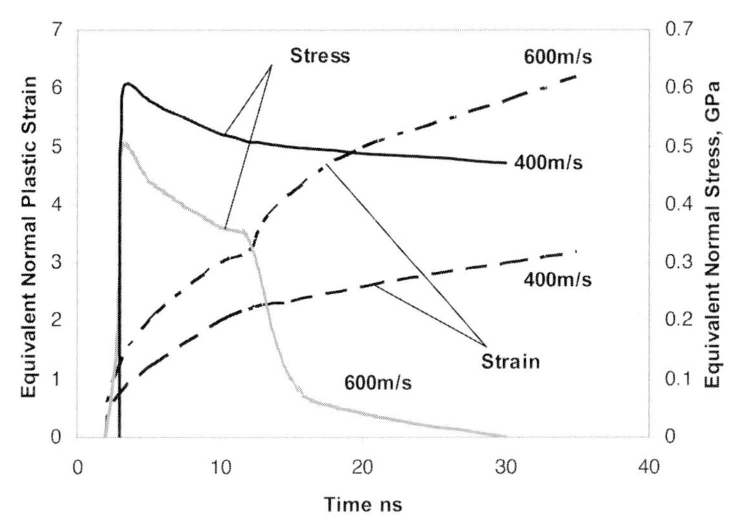

Fig. 2.9 Equivalent normal stress and strain, after Grujicic et al. (2004).

The empirical Johnson–Cook formula (2.17) is used in the computational analysis of one particle impact for determining the equivalent flow strength. In Eq. (2.17), A, B, and n represent the work-hardening constants, m represents the strain rate hardening constant, and ϑ is the thermal softening parameter. As Johnson and Cook have published tables of these parameters for many common materials, it is relatively simple to apply the flow law (2.17) in numerous one-dimensional deformation process computations with strain rates up to $10^6\ \mathrm{s}^{-1}$. These constants, however, need to be adequately defined for high impact velocities. For example, the strain rate sensitivity for Al appears to increase dramatically for the strain rates higher which have been achieved in pressure-shear plate impact testing ($> 10^6\ \mathrm{s}^{-1}$, Wright 2002).

The parameters used in the Johnson and Cook formula are determined from dynamic stress–strain curves for tension and compression at nearly constant strain rates and temperatures. Strain rate sensitivity and thermal softening, however, have a powerful influence on material behavior during strain localization. For this reason, the application of these values in the analysis of particle deformation by impact with strain rates higher than $10^6\ \mathrm{s}^{-1}$ needs to be revaluated.

Thermal sensitivity, a, and strain rate sensitivity, m, are known to be defined as

$$a = -\frac{1}{\rho C_p}\frac{\partial s}{\partial T} \quad \text{and} \quad m = \frac{\partial \ln s}{\partial \ln \dot{\gamma}} \tag{2.18}$$

Here the thermal softening parameter, ϑ, in Eq. (2.17) is a function of thermal sensitivity, a.

From an engineering point of view, in order to analyze the influence of each material parameter on adiabatic shear band formation during high velocity particle impact, it seems to be reasonable to use the constitutive equation in the most simple form – the power law (Zhou et al. 2006, Daridon et al. 2004):

$$\tau(\gamma, \dot{\gamma}, T) = \tau_0 \left(\frac{\gamma}{\gamma_0}\right)^n \left(\frac{\dot{\gamma}}{\dot{\gamma}_0}\right)^m \left(\frac{T}{T_0}\right)^a \tag{2.19}$$

Here τ_0 is the yield shear stress of the material in a quasistatic simple shear test, n is the work hardening constant, m is the strain rate hardening constant, and $a < 0$ characterizes the thermal softening. Also, γ_0 is the strain at yield point for a shear strain rate of $\dot{\gamma} = 10^{-4}\ \mathrm{s}^{-1}$, $\dot{\gamma}$ is a reference shear rate, T_0 is a reference temperature, and T is the current temperature. A simplified relationship including thermal softening and power law strain-rate hardening has been presented by Zhou et al. (2006):

$$\tau = \tau_0 g(T)(1 + b\dot{\gamma}^p)^m \tag{2.20}$$

Here τ_0 is the static yield stress, $g(T)$ is a monotonically decreasing function, $b - 1$ is a reference strain rate, and m is the strain-rate sensitivity. For simplicity, strain hardening may be neglected (Nesterenko 1995). The plastic flow law may be derived from Eq. (2.20):

$$\dot{\gamma}^p = \frac{1}{b}\left\{\left[\frac{|\tau|}{\tau_0 g(T)}\right]^{1/m} - 1\right\} \tag{2.21}$$

Zhou et al. (2006) consider two kinds of thermal softening: linear, as used by Nesterenko (1995),

$$g_1(T) = 1 - \alpha T \tag{2.22}$$

and exponential softening, as used by DiLellio and Olmstead (1997):

$$g_2(T) = exp(-\alpha T) \tag{2.23}$$

where α^{-1} is a reference temperature. In Eq. (2.22), $\alpha^{-1} = T_m - T_r$, where T_m is the melting temperature and T_r is the room temperature.

The strain localization during particle impact is governed by strain-rate hardening and thermal softening as a consequence of self-heating; it is described by the strain localization parameter (Grujicic et al. 2004) as

$$SL = \left(-\frac{\partial^2 s/(\partial\epsilon^p \partial\epsilon^p)}{(\partial s/\partial\dot{\epsilon}^p)\, s}\right)_{s=s_{\max}} \tag{2.24}$$

where $s_{\max}$ denotes the maximum flow stress on the stress strain curve at a particular particle velocity. Unfortunately, the direct determination of the partial derivatives of Eq. (2.24) is not possible, even in the one-dimensional case.

However, adiabatic shear localization evaluation based on the finite element analysis suggests values of SL around 1.6×10^{-4} s/GPa (Grujicic et al. 2004). It is interesting to note that calculations made with the model described in Eqs. (2.12)–(2.17) result in comparable tendencies in strain localization for all materials and material combinations. It is the opinion of the authors, however, that due the absence of real data concerning the thermal and strain rate sensitivity constants this conclusion should not yet be fully accepted. Nevertheless, SL is considered to be a critical condition for the onset of adiabatic strain localization, ultimately allowing authors such as Grujicic et al. (2004) and Papyrin et al. (2002) to calculate the critical particle velocities that are required to attain this critical value of SL, and consequently to achieve successful GDS deposition (see Table 2.1). Critical particle velocities also depend (at least in part) on the fundamental model parameters such as $\dot{\epsilon}_0^p$, s_{init}, and t_c. This is due to

i. the uncertainty that exists in determining the constants;

ii. the development of adiabatic shear bands after instability onset; and

iii. various features specifically related to the shock consolidation of particles.

The numerical modeling of shear localization by DiLellio and Olmstead (1997, 2003) appears to justify this statement. Figure 2.10 depicts a simplified model of this numerical solution (DiLellio and Olmstead 2003) that has been adapted for the Johnson–Cook equation with brass, nickel, and an iron and aluminum alloy. All materials exhibit a pronounced drop in shear stress (Fig. 2.10(a)) for distinct shear strains (defined by dimensionless time). Thus, it can be seen that the onset of localization is achieved more readily for nickel than for the iron and aluminum alloy. It is interesting that the sudden increase in temperature is observed to take place at different (nondimensional) times for each metal (Fig. 2.10(b)). Further, the values of nondimensional temperature parameters differ considerably. In fact, the temperature jump for iron is three times that of the aluminum alloy. The reason for this is not yet fully understood.

The dependence of the shear strain rate on nondimensional time is shown in Fig. 2.10(c). These numerical results suggest a strong relationship between the specific material and the onset of localization. The numerical scheme put forth by DiLellio and Olmstead (1997, 2003) provides an efficient approach by which to investigate shear localization in materials which obey the Johnson–Cook flow law (2.17). Unfortunately, as the simplified Johnson–Cook flow law neglects hardening, it is not without limitation. For this reason the numerical results that are obtained cannot be quantitatively compared to experiments; qualitative comparison of various materials is, on the other hand, quite ap-

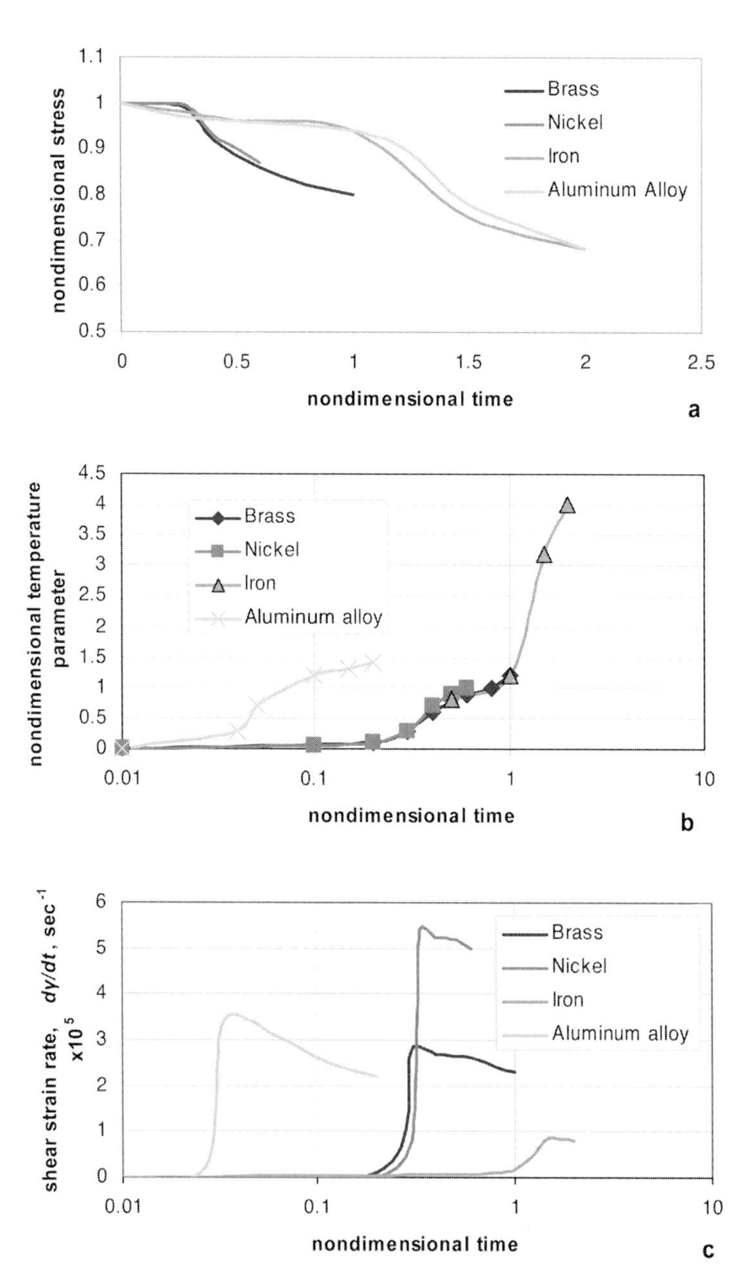

Fig. 2.10 Numerical solutions for the main parameters of adiabatic shear band evolution for various materials, after DiLellio and Olmstead (2003). (a) Nondimensional stress; (b) temperature parameter; (c) shear strain rate.

propriate (DiLellio and Olmstead 2003). This approach, in contrast to works of Assadi et al. (2003) and Grujicic et al. (2004), clearly illustrates the strong influence material type (and consequently strain rate and thermal sensitivity) has on the parameters of localization.

Assadi et al. (2003) and Grujicic et al. (2004) have determined the critical velocities of various metals (Table 2.1). In particular, Assadi et al. (2003) used an experimental method to determine the critical velocity which is detailed by Stoltenhoff et al. (2002). The procedure was based on the combination of calculating the particle velocity distribution using computational fluid dynamics (CFD) and measuring the corresponding mean velocity by laser Doppler anemometry (LDA). As a result the calculated velocity for the largest and slowest bonded particle was taken as the critical velocity for bonding. Two distinctive features of this method should be noted:

1. The critical velocity for one particle may characterize the GDS process for low particle concentrations in powder-laden jet only when this does not influence the particle consolidation process.

2. The GDS technology parameters of the experiment are not directly related to sprayed material characteristics. Additionally, the properties of the powder materials have not been accounted for.

It is therefore necessary to define the powder material characteristics that ultimately determine the overall effectiveness of the spraying process – this will undoubtedly result in deeper understanding and more effective process modeling. Additionally, the effect of particle concentration on the deposition efficiency should be studied in order to clarify the nature of powder consolidation in GDS.

2.2.3 Experiments Relating to Particle Impact

It has been proposed by Dykhuizen et al. (1999) that large deformations and pressures are required to achieve intimate contact between clean metal surfaces. In GDS, the bonding process is very similar to the explosive welding process (Crossland 1982). It is generally accepted that the formation of a solid-state jet of metal at the impact point between two metal planes promotes good bonding in explosive welding. This solid-state jet is responsible for breaking up surface layers, in turn causing better contact between the two metal plates. Examination of the craters that result from the GDS process also yields evidence of jet formation (Dykhuizen et al. 1999). In particular, the crater lips appear sharper in craters where the particle has bonded, whereas empty craters appear smother. The suggestion by Dykhuizen et al. (1999) that the sharper lip is an indication of large material deformations is reasonable. It remains of

Tab. 2.1 Evaluation of the critical particle velocities reported by Assadi et al. (2003) and Grujicic et al. (2004).

Particle material	Substrate material	Threshold particle velocity (m/s) Assadi et al.	Finite element analysis	SL factor analysis
Copper	Copper	570–580	575–585	571
Aluminum	Aluminum	760–770	760–770	766
Nickel	Nickel	600–610	620–630	634
316L	316L	600–610	620–630	617
Titanium	Titanium	670–680c	650–670	657
Copper	Aluminum	N/A	510–530	507
Aluminum	Copper	N/A	600–630	634
Copper	Nickel	N/A	570–580	571
Nickel	Copper	N/A	570–580	576
Copper	316L	N/A	570–580	574
316L	Copper	N/A	570–580	573
Copper	Titanium	N/A	520–550	514
Titanium	Copper	N/A	570-590	582

great interest, however, to evaluate real particle strains that result from impact – it seems that this is a key element in understanding the mechanism of GDS.

Certainly, in order to model high-speed particle impact, one must obtain reliable data for the fundamental parameters of GDS process. This, however, may only be achieved through experiment, the results of which have been presented by Papyrin et al. (2002). In this work the authors publish data related to the interaction of individual particles with substrates at various particle velocities. Here the particle velocities were altered through the application of various gas carriers, and measured by means of a laser Doppler anemometer and track methods. As this study of the individual particles included statistical data processing, accurate particle deformation results were achieved that can readily be applied to process modeling techniques. Some of these results are discussed below.

The dimensional theory analysis of particle deformation at normal impact on a surface has shown that the main parameters describing particle strain are $\rho_p V_p^2/H_p$, ρ_p/ρ_s, and H_p/H_s, where ρ_p, H_p, ρ_s, and H_s are the density and dynamic hardness of the particle and substrate, respectively. Experiments have shown that the ratios ρ_p/ρ_s and H_p/H_s do not have a significant effect on particle strain. In fact, the main GDS parameter is $\rho_p V_p^2/H_p$. The total particle strain may be estimated by $\epsilon_p = 1 - h_p/d_p$, where h_p is the height of the adhered particle and d_p is its initial diameter. Papyrin et al. (2002) have calculated the total particle strain from the following empirical equation, with

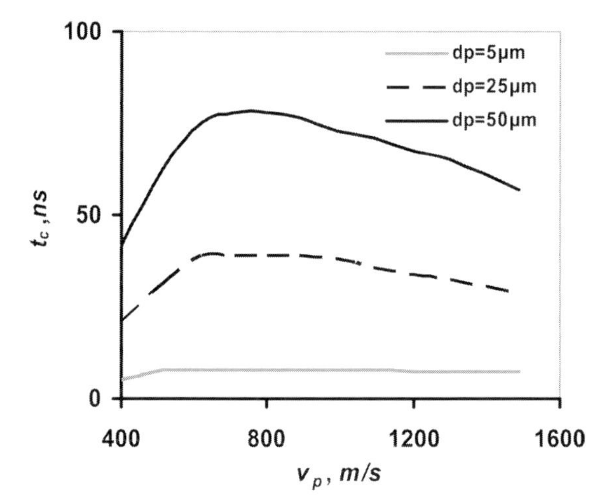

Fig. 2.11 Contact time versus the impact velocity, after Papyrin et al. (2002).

$H_p = 6 \times 10^7$ Pa and $k = 1.4$:

$$\epsilon_p = \exp\left(-k\frac{H_p}{\rho_p V_p^2}\right) \tag{2.25}$$

The particle strain rate parameters are also of great importance when characterizing the nature of GDS structure formation. For this reason it is necessary to determine the contact time during which particle plastic flow occurs. Contact time, t_c, is formally described as the time during which the impact particle decelerates from V_p to $V_p = 0$. Mathematically this may be approximated as

$$t_c \approx \frac{2\epsilon_p d_p}{V_p} \tag{2.26}$$

One can see from Eq. (2.26) that an increase in the particle size results in a proportional increase in the contact time. Some calculated contact times for aluminum particles having various impact velocities and diameters are shown in Fig. 2.11; typical values are in the range of 10–60 ns. This duration of time is sufficient for adiabatic shear band formation at the interface (see Figs. 2.7 and 2.9) for the known critical velocities.

Analysis of the particle shape cross-section after impact reveals that it can be approximated by a paraboloid of revolution, as is the case of uniform particle deformation (Fig. 2.12(a)) in spite of the adiabatic shear localization (Fig. 2.12(b)). In fact, the particle shape distortion due to particle impingement may be characterized by the relative contact radius, $\delta = r/r_{\text{final}}$, and the relative contact time $\tau = t/t_c$, where r_{final} is the final contact radius. The results of the

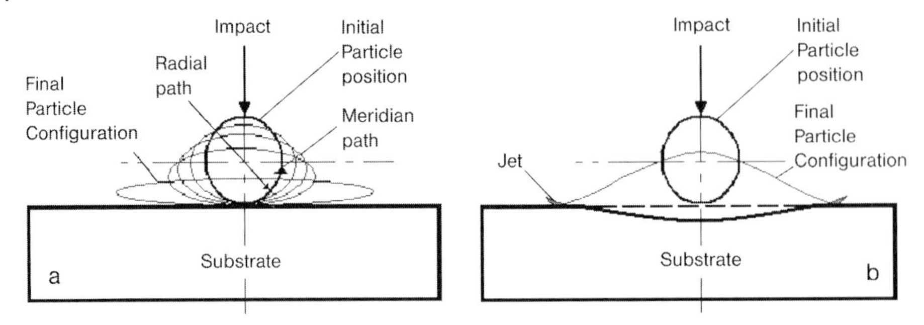

Fig. 2.12 One particle impact: (a) uniform particle deformation, (b) localization of deformation.

experiments by Papyrin et al. (2002) (Fig. 2.13) reveal that the relative contact radius can be approximated by the equation $\delta = \tau^{1/3}$ for particle velocities ranging from $v_p = 500$ to 1000 m/s and by $\delta = \tau^{1/2}$ for particle velocities ranging from $v_p = 1000$ to 1500 m/s. One can also see that large differences in particle velocities do not result in changes in the approximation equations. This is due to the significant adiabatic shear localization of deformation that takes place at the interface (Fig. 2.12(b)). This process leads to the rapid radial flow of metal near the interface, where the specific radial velocities may be calculated at various distances from the particle center. Figure 2.13(b) depicts the results of radial velocity calculations at $t = 20$ ns for three distinct regions within the particle. It is obvious that the radial velocity reaches a maximum in the near interface region with some nodes at $r = 30$ μm achieving velocities of approximately 1200 m/s. However, the radial velocity of the nodes at $r \approx 0$ is not large enough to create the adiabatic shear band (Fig. 2.13(b)). Radial velocities which are able to achieve the critical shear strain rates that are necessary for bonding are shaded in Fig. 2.14. It can be concluded from these data that shear localization during particle impact does not ensure uniform bonding conditions at the interface. The thickness of layers having maximal radial velocity was estimated to be $0.05d_p$. This result is in reasonable agreement with the work by Assadi et al. (2003), who modeled the flow stress and strain distribution in the near wall region (shown in Fig. 2.15). It is believed that this layer thickness corresponds to that of the adiabatic shear band, with severe shear strains and high shear strain rates occurring in this region. Thus, the determination of adiabatic shear band thicknesses for particle impacts within the GDS process serves to define the shear localization parameters for each case of interest.

Analysis of the modeling results for impact velocities of 300 m/s (Fig. 2.15) reveals a monotonic increase in equivalent stress in the vicinity of the interface. This is due to the hardening effects that occur as a result of high strains and strain rates at the interface. On the other hand, the high degree of plastic-

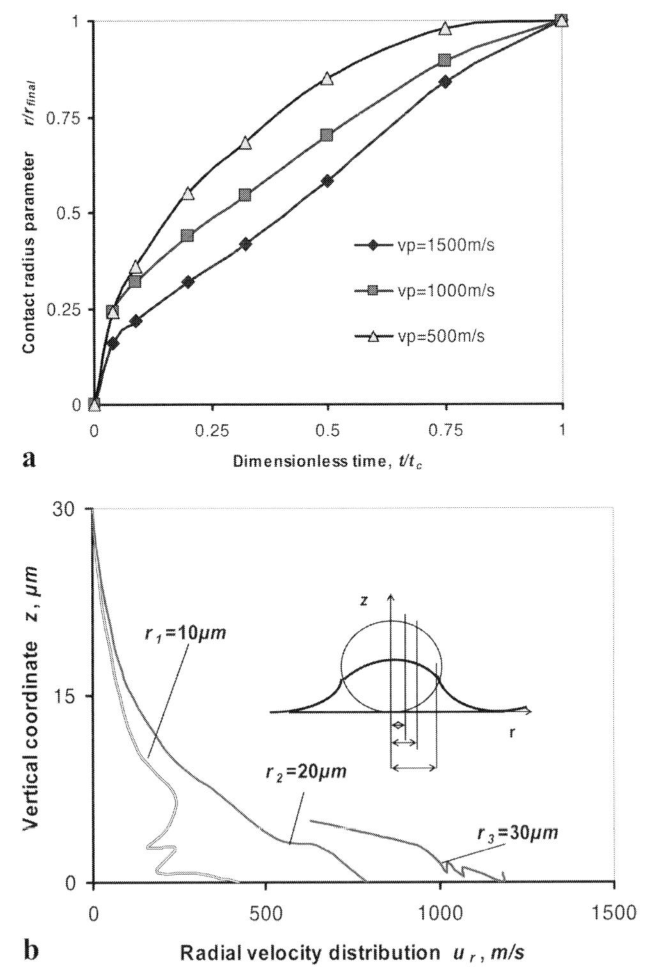

Fig. 2.13 Relative contact radius versus nondimensional contact time at the different particle velocities. (a) Numerical modeling results for the impact of an Al particle ($d_p = 60$ μm) with velocity 800 m/s and temperature 300 K on a rigid substrate. (b) Plastic flow of particle nodes in the radial direction during impact at the time of 20 ns in three sections of the particle (friction at the interface is zero), after Papyrin et al. (2002).

ity at the interface can result in softening due to adiabatic heating. At higher particle velocities this softening has been shown to exceed the hardening effects, leading to adiabatic shear instability at the interface. This effect may be clearly observed for impact velocities of 900 m/s. For these velocities, despite the high magnitude of plastic strain at the interface, the shear strength of the material falls to near zero values due to dominance of thermal softening. Thus, thermal softening and adiabatic shear instability at the interface can

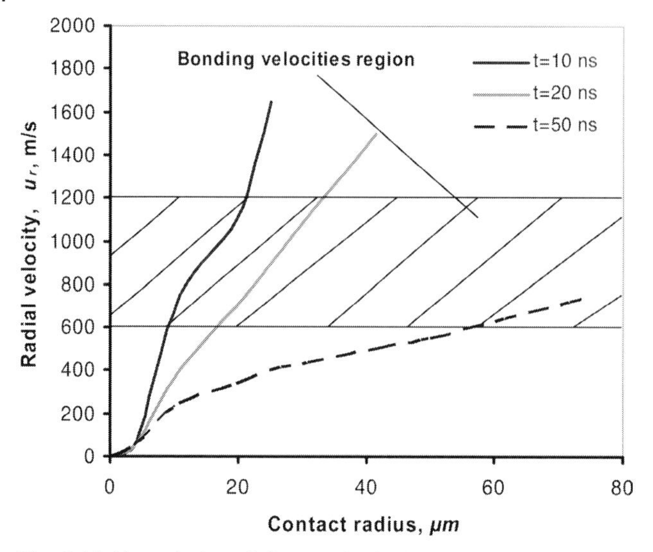

Fig. 2.14 Numerical modeling results for the impact of an Al particle, after Papyrin et al. (2002).

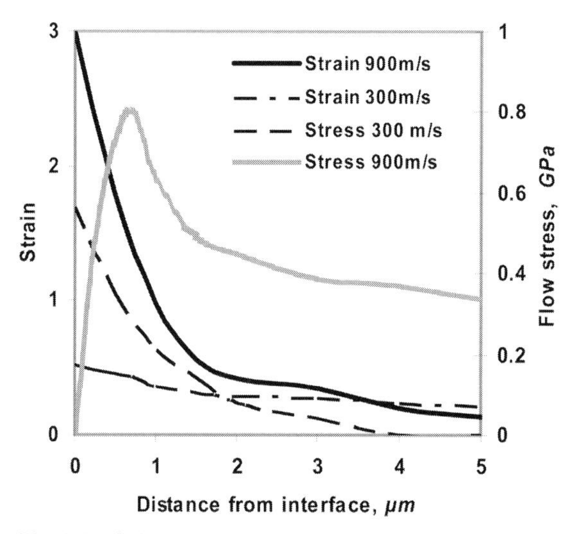

Fig. 2.15 Calculated profiles of plastic strain and flow stress, modeled after Assadi et al. (2003).

play significant roles within the collection of microscopic processes that necessarily lead to the successful bonding of particles in GDS. However, the lack of experimental data for material constants at high particle velocities limits the accuracy of numerical models in characterizing real GDS processes.

2.3 Concluding Remarks

In this chapter, the impact parameters of a single particle have been described in a qualitative way. Further, the principal experimental and theoretical modeling methods that are commonly used to examine this process have been briefly reviewed. The main points that must be carried forward to the following chapters are that shear localization and adiabatic shear band formation appear to have a profound effect on the bonding process. Also, it is of great importance to effectively estimate the strain rate and thermal sensitivity of each substrate for various impact velocities. Finally, the role of the initial yield strength and thermal softening processes in adiabatic shear band formation during particle impact has to be further evaluated for both classical and low pressure GDS. It is only through this analysis that analytical materials models may be used to help interpret results and, in turn, develop an effective approach that permits the accurate calculation of specific GDS parameters.

3 Densification and Structure Formation of the Particulate Ensemble

3.1 Identification of Various Phenomena

The main processing characteristics of densification and bonding structure formation in GDS are considered to be similar to those in processes such as explosive welding or shock-wave powder compaction (Assadi et al. 2003, Schmidt et al. 2003, 2006). In all of these processes, bonding is achieved only when a critical impact velocity is exceeded. The critical velocity in these processes is generally perceived to be a function of the material properties (Schmidt et al. 2006).

The most important microstructure formation processes involved in gas dynamic spraying have to be identified and discussed through an energy approach. The primary feature common to both GDS and shock consolidation processes is that energy is dissipated via a shock wave. Interparticle adiabatic shear band formation, vorticity, voids, and particle fracture may also occur due to the plastic deformation. Various energy dissipation processes take place: plastic deformation, interparticle friction, microkinetic energy, and defect generation. Meyers et al. (1999) have introduced an analytical expression which may be used to determine the energy that is necessary to consolidate a powder, as a function of strength, size, porosity, and temperature, based on the materials properties. This formula allows one to predict the pressure that is required to shock consolidate materials, as well to characterize the role of each energy dissipation processes. Based on the utility of this expression, it is therefore beneficial to organize the GDS microstructure formation processes in accordance with the formalism presented by Meyers et al. (1999).

The proper identification and quantitative evaluation of the various phenomena that occur during the impingement of a powder ensemble upon a substrate is necessary to estimate the overall energy requirements for forming powder coatings. Figure 3.1 provides a schematic representation of these phenomena for GDS. There are in fact two groups of processes that must be characterized: (1) impact processes with energy – these can be computed from phenomenological parameters such as particle velocity, yield stress, material density, etc.; and (2) structure formation processes which define the energy

Introduction to Low Pressure Gas Dynamic Spray Roman Gr. Maev and Volf Leshchynsky

ISBN: 978-3-527-40659-3

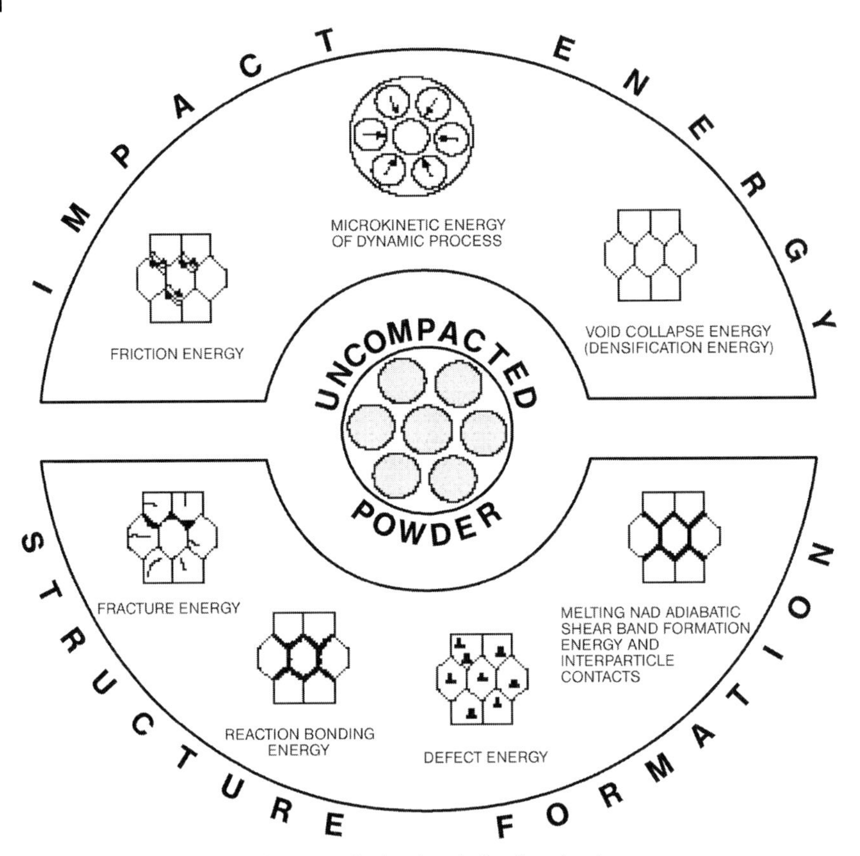

Fig. 3.1 Various models of energy dissipation during the shock compression of powders, after Meyers et al. (1999).

dissipation. The energy for each process may be computed upon determining the structural parameters of the system.

Group 1: impact processes with energy

a. *Plastic deformation energy.* The powder material is being plastically deformed upon impact, resulting in the collapse of the voids due to the plastic flow of each particle, as shown in Fig. 2.12(a).
b. *Microkinetic energy.* The localized plastic flow of the material (Fig. 2.15(b)) is achieved by means of the dynamic features of the shock consolidation process. This leads to interparticle impacts, friction, and plastic flow beyond which it is geometrically necessary to collapse the voids, as described in (a). This component is referred to as "microkinetic energy" (Nesterenko 1995), a process of intense energy dissipation with localized heating. Microkinetic energy results from particle kinetic energy inducing the formation of a jet at the particle–substrate (or particle–

particle) interface, in turn bringing about an increased temperature (Fig. 2.12(b)), as observed in the GDS process. The entire plastic deformation path is replaced by deformation in a localized area. However, the kinetic energy acquired by the material elements (being plastically deformed) eventually dissipates into thermal energy, as in conventional plastic flow.

c. *Friction energy.* The relative movement of particles at the shock front as a result of the applied stress brings about interparticle friction – this plays a significant role in energy deposition. Both stress and deformation within granular materials are known to be nonuniform, particularly at the microscale of particle groups (Molinari 1997). The various material parameters (or factors such as the coefficient of friction) that yield stress, cohesion loading, etc. influence the characteristics of particle ensemble flow. The high friction forces that can be observed at the particle interface are due to particle flow and high radial velocities (Fig. 2.14).

Group 2: structure formation processes

d. *Adiabatic shear band formation and melting at the particle–substrate or particle–particle interface.* Although the intensive shear strains that are localized at the interface may eventually lead to material melting, with GDS this is not usually the case.
e. *Defect energy.* Quantitative assessments of defect concentrations in shock-wave deformation reveal this factor to have only a minor influence on total energy losses (Meyers et al. 1999).
f. *Fracture energy.* Brittle particles such as oxides and carbides may fracture due to impact. The comminuted particles can more efficiently fill the voids.
g. *Shock initiated chemical reactions.* Some chemical reaction at the particle–particle interface may occur as a result of powder mixtures being consolidated during impact. In the case of exothermic reactions, this process contributes additional energy to the powder surface and acts to facilitate bonding.

These GDS impact processes, although they may be conveniently categorized into the distinct groups outlined above, are not entirely independent. Indeed, as presented in Fig. 3.1, there exists some degree of overlap. For example, as the transformation from uniform plastic flow (Fig. 2.12(a)) to localized flow (Fig. 2.12(b)) is difficult to define, microkinetic energy and plastic deformation energy are tightly connected. On the other hand, fracture energy and plastic deformation energy are competing processes – densification can occur as a result of either one or a combination of both processes (Meyers et al. 1999). As presented above, however, the simplified explanation of these energy components allows one to more distinctly define the role of each process in GDS structure formation.

3.2 Observations of GDS Consolidated Materials

In order to effectively define the general trends in powder coating formation it is first necessary to review the essential features of high and low-pressure GDS sprayed materials. To do this it is sufficient to analyze the structures of both Al and Al-based coatings. The characteristics of Al coatings spayed with average air gas temperatures of 204°C and pressures of 2.4 MPa were analyzed in detail by Van Steenkiste et al. (2003) for relatively large particles (a kinetic spray process). Some of these results are shown in Figs. 3.2 and 3.3. Except at the points of contact, one can see a small degree of internal grain distortion. The bonding points are indicated with arrows (Fig. 3.2(a)). An SEM image of the coating at higher magnification (Fig. 3.2(b)) reveals the internal grain structure of the particles, as well as the intensive plastic flow in the area of bonding. Figure 3.2 clearly shows the lack of deformation that can occur for the same particles at locations sufficiently removed from the contact area. The optical photograph of the etched coating (Fig. 3.3) shows similar patterns of grain distortion that are localized in the contact area (again denoted by an arrow). The particle flattening that occurs due to high-pressure GDS impact is known in the range of $d_r/d_z = 3$–8 (Fig. 2.14). Indeed it is shown to be more intensive for particle–substrate impingement. The individual particle impact behavior seems to govern the process of bonding for this case, and is dependent on the concentration of particles in the near substrate region.

Results similar to Van Steenkiste et al. (2003) were obtained by Morgan et al. (2004). Here the cold spray process proved to be an effective means for bonding aluminum particles through high kinetic energy impacts which resulted in the plastic deformation. This bonding was found to take place without significant material heating. In fact, the deposits analyzed in this study exhibited occasional material continuation across the particle interfaces, suggesting sufficient kinetic energy to break the surface films and generate metallurgic bonding. Thus, the predominant mechanism for bonding appears to be mechanical interlocking. Further, the analysis of the fracture surface reveals that the observed interlocking of particles occurs in a laminar manner, with each particle conformal to the predeposited material. It is further suggested that an increase in the particle velocity may bring about increased bonding.

The existence of jets or vortex formations due to GDS impact has been observed in separate cases. For example, Fig. 3.4 depicts the formation of vortexes and jets during the high-pressure GDS of Cu powder (Champagne et al. 2005).

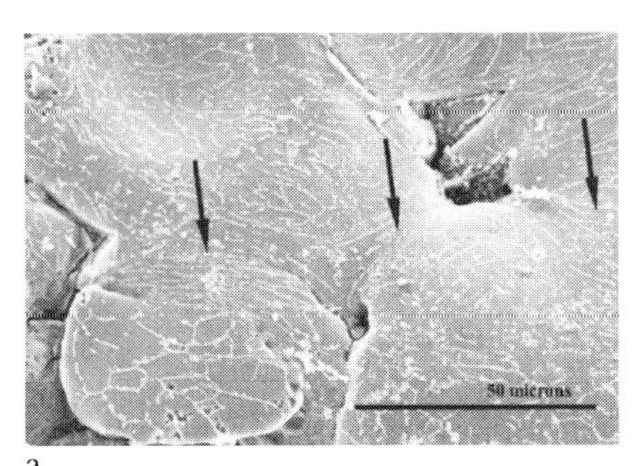

a

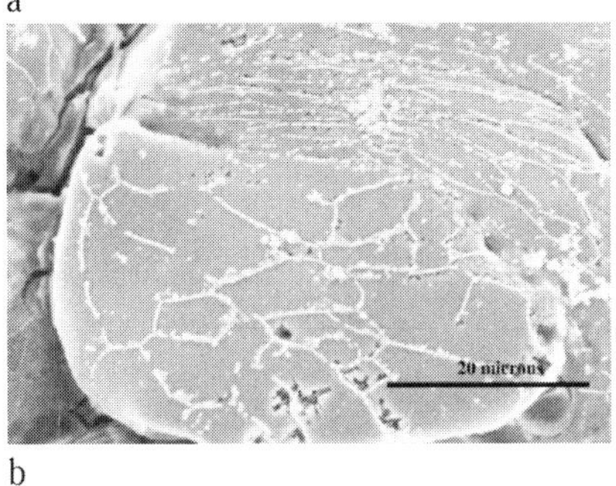

b

Fig. 3.2 (a) SEM image of an electropolished sample of a coating cross-section produced at an air carrier gas temperature of 260°C. (b) Magnified view of the electropolished particle shown in (a). Here the internal plastic flow of the grain boundaries at the impact site and the preservation of the original grain structure in locations removed from the impact area are shown. (Courtesy of Van Steenkviste et al. 2002)

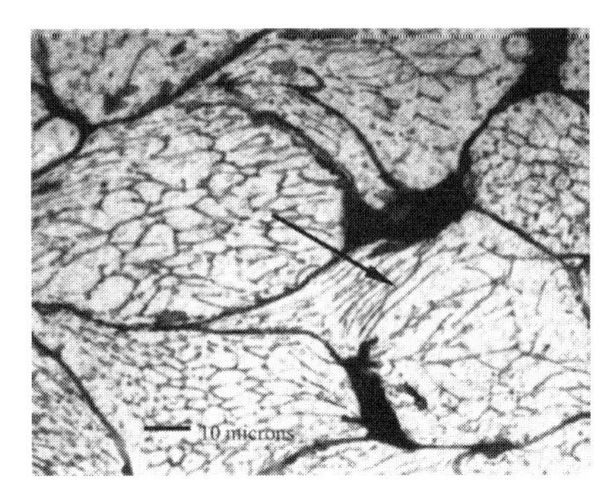

Fig. 3.3 Photographs of the air carrier gas at a temperature of 315°C (600°F), as described in the text (magnification 1000×). (Courtesy of Van Steenkiste et al. 2002)

An examination of the coating structure in the low-pressure GDS case was undertaken for Al–Zn–Ti and Cu–Zn–W coatings (Figs. 3.5–3.7), whose composition and parameters are presented in Table 3.1. The deformed Al particles are shown in Fig. 3.5 (Al-based powder mixture), with the flow of particles

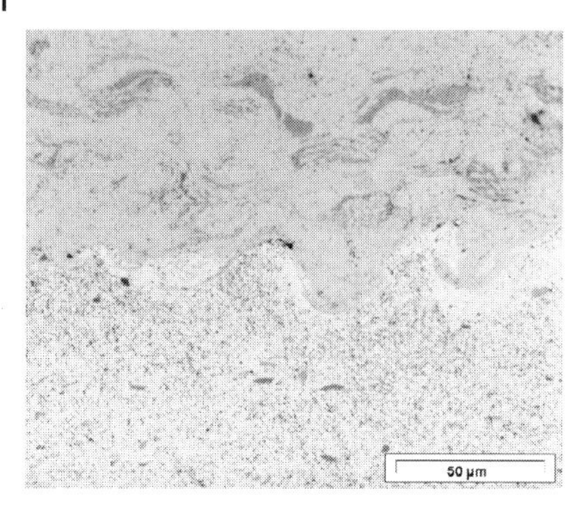

Fig. 3.4 Copper–aluminum interface (unetched). (Courtesy of Champagne et al. 2005)

due to impingement indicated using arrow B. Some local regions of Al particle bonding are shown by arrows C, while the bonding of the Al and Zn particles are shown by arrows D. The densification of Zn particles is indeed quite good, resulting in the collapse of the voids. Some deformed Zn particles are shown by arrow E. The overall kinetic energy of the Zn particles is small due to their lesser average particle diameter (∼ 10 μm). However, the intensive compaction of Zn agglomerates is facilitated by the larger Al and Ti particles (∼ 50 μm), having high kinetic energy, which act as penetrators. The alignment of the deformed particles does not coincide with the direction of impact (shown with arrow A) resulting in severe shear strains at the interparticle contacts due to inclined impact.

Tab. 3.1 Powder compositions for low-pressure GDS.

Recipe name	Composition (vol.%)					Air pressure (MPa)	Air temperature (°C)
	Al	Zn	W	Ti	Cu		
Al-based	50	25	–	25	–	0.5	200
Zn-based	10	65	–	25	–	0.5	200
Cu-based	–	25	–	25	50	0.6	450

GDS deposition of Zn-based powder mixtures leads to a change in the mechanism of coating structure formation (Fig. 3.6). In this case the main process at the substrate-coating interface seems to be the compaction of Zn particles (shown with arrow B). Ti (arrow E) and Al particles (arrow D) play the role of penetrators, as in the previous case, creating a compaction effect that results in a high density of Zn particle agglomerates (arrow C). It is difficult, however, to achieve stable compaction in all regions of the powder layer due to a high content of Zn particles with low kinetic energy. For this reason some areas with high porosity are observed (Fig. 3.6(b)). Additionally, the dis-

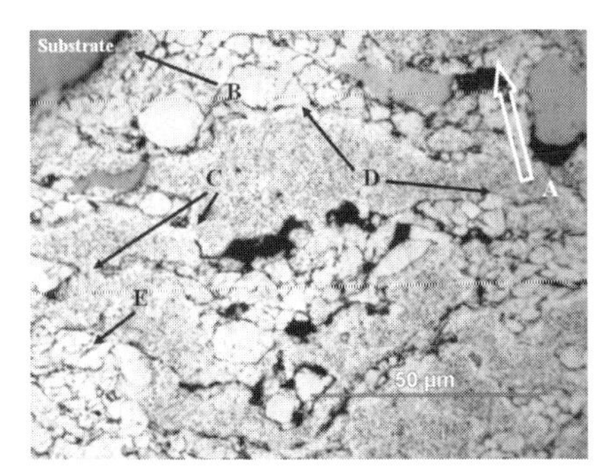

Fig. 3.5 LPGDS of a 50Al–25Zn–25Ti (wt.%) powder mixture. Air temperature, 300°C; air pressure, 6 bar.

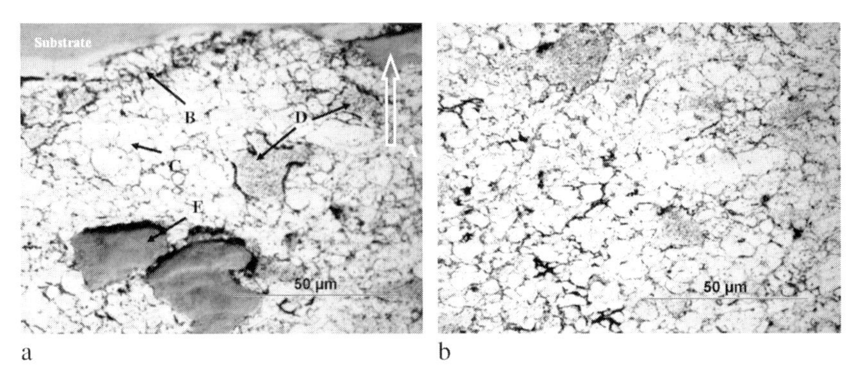

Fig. 3.6 LPGDS of a 10Al–65Zn–25Ti (wt.%) powder mixture. Air temperature, 300°C; air pressure, 6 bar. (a) Substrate area; (b) top layer.

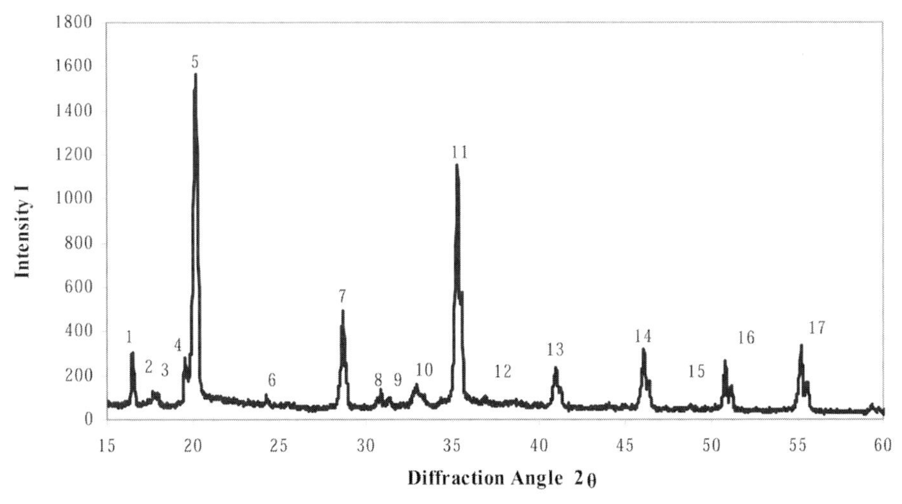

Fig. 3.7 X-ray spectra of a Ti-50%,Al-10%,Zn-40% LPGDS coating. Source of the X-rays – Mo (45 kV). 1,4,6,7,14,15,17 – Ti_3Al; 2,9,16 – AlTi; 3,5,8,12,13 – $Zn_{0.68}Ti_{0.32}$.

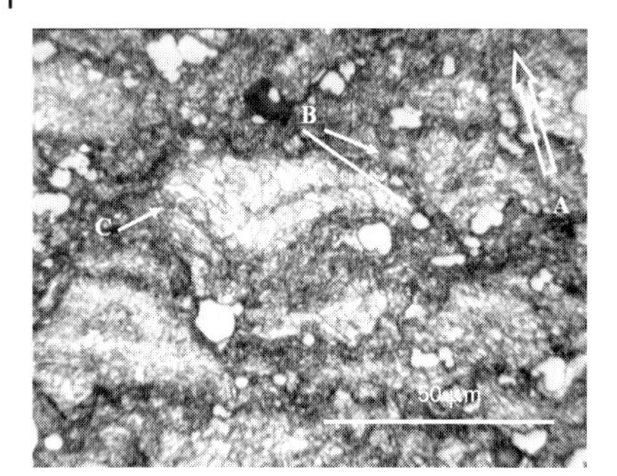

Fig. 3.8 LPGDS of 50Cu–25W–25Zn (wt.%) powder mixture. Air temperature, 400°C; air pressure, 6 bar (etched).

tortion of Al particles is very small in comparison with the Al-based mixture (Fig. 3.5). This is evidence of the low yield stress of Zn, which is easily deformed by the Al and Ti particles. Thus, the shock compaction feature seems to be most consequential in the low-pressure GDS deposition of powder mixtures, as compared with high-pressure GDS processes.

One cannot see the effects of jet formation in the case of low-pressure GDS of the Al–Zn–Ti powder mixtures. This, however, does not imply that localization of deformation and adiabatic shear band formation does not occur. As shown by Van Steenkiste et al. (2003), the plastic deformation results in the creation of bonding sites (shown by the arrows in Figs. 3.2 and 3.3). Additionally, localization of deformation results in reaction bonding. In fact, due to this phenomenon, some new phases are generated that may be clearly discerned in X-ray spectra data (Fig. 3.7). The peaks of the Ti_3Al, AlTi, and $Zn_{0.68}Ti_{0.32}$ compounds are indications of reactions at points of interparticle contact.

On the other hand, jet formation is observed in the case of Cu–Zn–W powder mixture deposition by low-pressure GDS. Evidence for vorticity, or redundant plastic deformation, is presented in Figs. 3.4 and 3.8, denoted by arrows B. The configuration of the Cu particles is similar to those in Fig. 2.12(b). In fact, the grain structure of the Cu particles indicates that some small grains have arisen in the area of interparticle contact (arrow C). These clear regions are microcrystalline or nanocrystalline in structure. The tungsten particles enhance the deformation and bonding effects due to high velocity impact. It is also important to note the voids in Figs. 3.5, 3.6, and 3.8; these contribute to the embrittlement of the shock-consolidated material because they act as initiation sites for cracks during deformation.

3.3 Energy Requirements for GDS Shock Consolidation

In accordance with the previously described process classification scheme (Fig. 3.1), an energy calculation may be made for each component of energy. For the first group of processes, the energy requirements may be calculated based on the phenomenological impact parameters (i.e., the particle velocity and the basic characteristics of the powder layer such as current density, particle diameter, coefficient of interparticle friction, etc.).

3.3.1 Plastic Deformation Energy

The plastic deformation energy (or void collapse energy) is strongly affected by particle geometry, particle contact area, and the mechanical relationship between adjacent particles (Meyers et al. 1999). A review of relevant models of densification has been published by Olevsky et al. (1998). In accordance with the model presented by Helle et al. (1985), the interparticle pressure, P_{int}, is found to be related to the effective applied pressure, P_{eff}, by

$$P_{\text{int}} = \frac{P_{\text{eff}}(1-\rho_0)}{\rho^2(\rho-\rho_0)} \tag{3.1}$$

where P_{eff} and ρ are the applied stress and current relative density, respectively, and ρ_0 is the initial relative density. Taking into account the plastic yield condition,

$$P_{\text{int}} \geq 2.70 Y_y \tag{3.2}$$

where Y_y is the flow stress of the material and P_{eff} may be estimated by

$$P_{\text{eff}} = P_y = 2.97\,\rho^2\,\frac{\rho-\rho_0}{1-\rho_0}\,Y_y \tag{3.3}$$

The specific energy (per unit volume) required to densified the material is given by

$$E_{\text{vc}} = -\int_{\rho_0}^{\rho} P_y\,\frac{1}{\rho}\,d\rho \tag{3.4}$$

resulting in the following energy requirements for void collapse:

$$E_{\text{vc}} = \frac{2.97 Y_y}{1-\rho_0}\left[\rho\rho_o - \frac{(\rho^2-\rho_0^2)}{2}\right] \tag{3.5}$$

Another commonly used constitutive model has been presented by Carroll and Holt (1972). A comparison between the calculated pressures (as functions of porosity) for both models is shown in Fig. 3.10. The Helle et al. (1985) model represents both the initial and intermediate stages of consolidation (up to $\rho = 0.92$), whereas the Carroll–Holt model more effectively illustrates the collapse of isolated spherical pores (in the last stages, from $\rho = 0.92$ to 1).

3.3.2
Microkinetic Energy

Nesterenko (1995) developed a model which describes the relative movement of particles under dynamic compression, as shown in Fig. 3.11. This is a modification of the Carroll and Holt (1972) model which considers the porous material as a hollow sphere which is pressurized from the outside and, consequently, undergoes changes in its internal radius. The Nesterenko (1995) modification of the Carroll–Holt model enables the introduction of a size scale and separation of the plastic deformation process into two components, one of which has significant microkinetic energy. The external shell, with a mass m, is directly responsible for the microkinetic energy, since it impacts the internal core, which is considered to be stationary. In this model, the external shell impacts the central core (with radius c) at an impact velocity V. This impact velocity is approximated as the ratio of the average pore volume $(a_0 - c)$ to the contact time. The representative element for the GDS consolation of a bimodal powder mixture, as depicted in Fig. 3.10, may be transferred to the Nesterenko (1995) model (Fig. 3.11). The impact velocity for this model is given by $V_N = (a_0 - c)/t$, where t is the contact time. The specific microkinetic energy (per unit volume) is given as

$$E_k = \frac{1}{2} m V_N^2 = \frac{1}{2} m \left(\frac{a_0 - c}{\tau} \right)^2 \tag{3.6}$$

where m is a volumetric mass of the external shell, which, according to Nesterenko (1995), is equal to

$$m = \frac{b_0^3 - a_0^3}{b_0^3 - a_0^3 + c^3} \rho_T \tag{3.7}$$

Here ρ_T is the theoretical density of the material. In the case of the GDS model with a bimodal powder mixture (Fig. 3.11), the mass of the external core is defined by the geometry of a representative element.

The kinetic energy E_p transferred to the powder layer that is to be densified is defined by the velocity of the incident particle, V_p, and its mass, m_p. The contact time may be estimated by Eq. (2.26). A porous layer with diameter D and initial height H_0, having a relative density of $\rho_0 = 0.6$–0.7 may be densified by a one-dimensional strain (cylinder with sizes of $D =$ const and $H = HT$) up to the relative theoretical density $\rho_T = 1$. The strain of the porous layer is

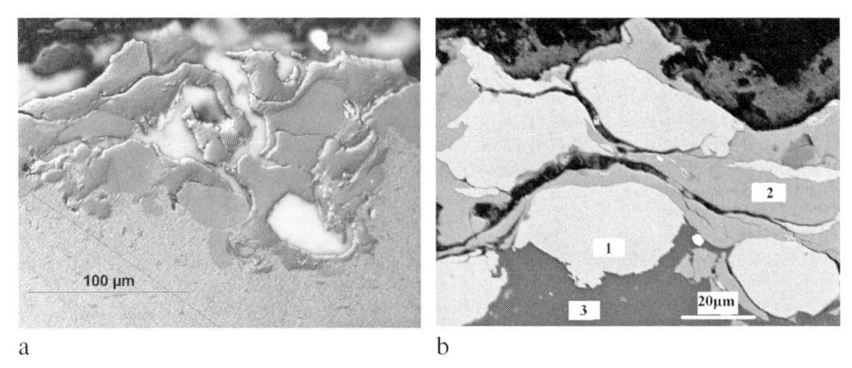

Fig. 3.9 Super deep penetration of Inconel particles (1) into the Al substrate, (3) at low-pressure GDS. Particle velocity 520 m/s. Jet formation of Ti particles (2). (a) Light microscopy image; (b) SEM image.

$H_T = \rho_0 H_0$, and the impact velocity of porous layer v_L is

$$v_L = \frac{\Delta H}{t_c} = \frac{1}{t_c}(H_0 - \rho_0 H_0) \tag{3.8}$$

If the entire kinetic energy of the incident particle (Fig. 3.9) is transformed to microkinetic energy, the kinetic energy of the external core, E_{core}, in the Nesterenko (1995) model may be calculated as

$$E_{\text{core}} = \frac{m_{\text{core}} v_L^2}{2} \tag{3.9}$$

The mass of the external core may then be calculated as

$$m_{\text{core}} = \frac{m_p\, v_p^2\, t_c^2}{H_0(1 - \rho_0)^2} \tag{3.10}$$

3.3.3
Frictional Energy

In accordance with Meyers et al. (1999), the frictional energy, E_f (based on pyramidal coordination and one-dimensional strain), may be calculated as follows:

$$E_f = f\mu\delta r = \frac{0.1P\xi Z\mu}{\rho} \tag{3.11}$$

where f is the applied force normal to the interparticle contact areas, μ is the friction coefficient at the interface, δr is the displacement at the interface, P is the stress that is applied to the powder layer (Fig. 3.10), ρ is the density, ξ

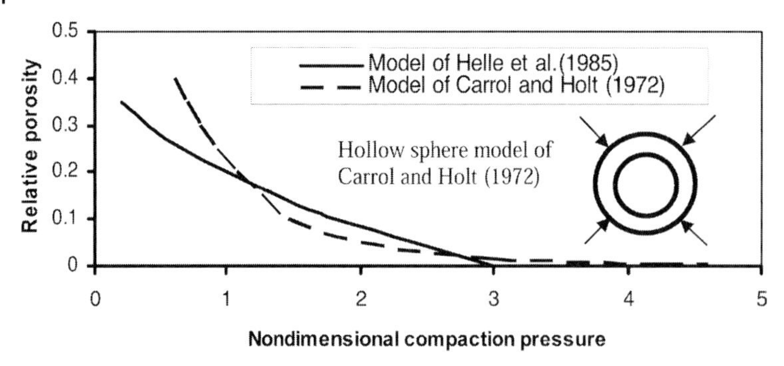

Fig. 3.10 Theoretical densification curves of a powder body.

is a fraction of the contact surface, and Z is the coordination number of the particles in the powder layer. These parameters are evaluated for a bimodal powder mixture GDS model.

For the second group of processes mentioned above, the energy requirement calculations may be based on the structure parameters and material constants of the sprayed coatings. As above, the energy losses in GDS consolidation may be characterized by the two primary structure formation processes: (d) adiabatic shear band formation and (e) generation of defects. The other processes of classical shock compaction such as melting or particle fracture are described by Meyers et al. (1999), although the first of these is not usually observed at GDS.

3.3.4 Adiabatic Shear Band Formation Energy

The specific energy of adiabatic shear band formation (per unit volume), E_{ASB}, can be generally expressed in the following form:

$$E_{\mathrm{ASB}} = \Theta_{\mathrm{ASB}} \left[\bar{C}_p (T_{\mathrm{ASB}} - T_0) + H_{\mathrm{ASB}} \right] \tag{3.12}$$

where Θ_{ASB} is the volume fraction of the adiabatic shear bands, $\bar{C}_p$ is the average value of specific heat per unit volume, T_{ASB} is the temperature of adiabatic shear bands formation, T_0 is the initial temperature of the powder, and H_{ASB} is the latent heat of adiabatic shear band formation per unit volume. The last parameter, H_{ASB}, is similar to the latent heat of melting, H_m. Also, based on one particle impact modeling experiments, H_{ASB} may be evaluated as a portion ψ of H_m, i.e., $H_{\mathrm{ASB}} = \psi H_m$.

3.3.5 Defect Energy

The shock energy stored in the elastic dislocation fields can be estimated from the specific energy of a dislocation line (per unit volume):

$$E_d = \left(\frac{Gb^2}{10} + \frac{Gb^2}{4\pi} \ln \frac{\rho_d^{1/2}}{5b} \right) \rho, \tag{3.13}$$

where G is the shear modulus, b is the Burgers vector, ρ_d is the dislocation density, and ρ is the particle density. The defect energy per unit mass, E_d/ρ_0 (where ρ_0 is the initial density of the particle), has been calculated and is found to be between 1 and 15 kJ/kg. This is a part of the deformation energy and is incorporated into the void collapse energy (Meyers et al. 1999).

3.4 Computation of ASB Energy Parameters

Energy balance within the particle consolidation processes plays an important role in GDS. As such we turn our attention to the Meyers et al. (1999) energy balance equation for GDS. The specific total shock energy can be calculated as a sum of the specific void collapse, microkinetic, and frictional energy. Hence,

$$E_T = E_{\text{VC}} + E_k + E_f \tag{3.14}$$

This energy, in turn, is equal to the sum of the thermal energy, E_t, and the defect energy, E_d. The defect energy – which has been found by Meyers et al. (1999) to be negligible – is a component of the plastic deformation energy. In GDS, thermal energy results in adiabatic shear band formation (E_{ASB}) and substrate heating (E_h) to a temperature of T_h. Thus the energy balance equation can be expressed as

$$E_t = E_{\text{VC}} + E_k + E_f = E_d + E_{\text{ASB}} + E_k \tag{3.15}$$

Therefore, the adiabatic shear band formation energy is a direct result of plastic deformation, microkinetic, and friction energies. Assuming that the contribution of the deformation energy into adiabatic shear band formation energy is negligible, we have

$$E_{\text{ASB}} + E_k = E_k + E_f = \frac{1}{2} \frac{m_p \, v_p^2 \, t_c^2}{H_0(1-\rho_0)^2} \left(\frac{a_0 - c}{\tau} \right) + \frac{0.1 P \xi Z \mu}{\rho} \tag{3.16}$$

The data of Meyers et al. (1999) imply that there exists a linear relationship between the pressure required for shock consolidation and the initial yield strength of the material. This indicates that the strength of the material has a significant influence on the consolidation process. This, in turn, supports the assumption that adiabatic shear band formation takes place after substantial densification has occurred. It can be deduced from the results shown in Fig. 2.14 that bonding may be achieved only for radial velocities that exceed 600 m/s. Therefore, particle deformations that occur for contact radii up to approximately 15 μm (for particles with diameters $r_p = 60$ μm) will do so without bonding. For this case the volume fraction of adiabatic shear bands Θ_{ASB} is estimated as

$$\Theta_{ASB} = 1 - \frac{3 - (r_f^2 - r_0^2)}{r_p^3} b_{ASB} \tag{3.17}$$

where r_f is the radius of the particle after the consolidation process (Fig. 2.14). Therefore, the ASB formation energy may be computed as

$$E_{ASB} = \left[\bar{C}_p(T_{ASB} - T_0 - T_k) + H_m\right] \times \left[1 - \frac{3(r_f^2 - r_0^2)}{r_p^3} b_{ASB}\right] \tag{3.18}$$

Here the term T_h was added to take into account the temperature rise that is assumed to result from the GDS consolidation process. Substituting Eq. (3.16) into (3.17) one obtains

$$\frac{1}{2}\frac{m_p\, v_p^2\, t_c^2}{H_0(1-\rho_0)^2}\left(\frac{a_0 - c}{\tau}\right) + \frac{0.1P\xi Z\mu}{\rho} = \left[\bar{C}_p(T_{ASB} - T_0 - T_k) + H_m\right] \times \Theta \tag{3.19}$$

Thus, the volume fraction of adiabatic shear bands, Θ_{ASB}, in GDS may be calculated through a combination of both theoretical modeling and experimental results.

3.5
Shear Localization During Particle Shock Consolidation

The actual mechanism by which solid particles deform and bond during the GDS process is still not well understood. Prevailing theories for GDS bonding postulate that solid particles undergo extensive plastic deformation during impact, disrupting thin surface films on the substrate, achieving intimate conformal contact with the target surface (Grujicic et al. 2003). This theory is supported by a number of experimental findings. In fact, while a wide range of ductile materials can be successfully cold-sprayed, nonductile materials such as ceramics must be cold sprayed in a powder mixture with ductile materials. Also, the fact that the particle kinetic energy at impact is typically much lower than the energy required to melt the particle further suggests that the deposition mechanism is a solid-state process. Thus, the kinetic energy of an incoming particle must be completely converted into heat and strain energy within the coating and substrate. The lack of melting is directly confirmed through the micrographic examination of coated materials (Grujicic et al. 2003). Finally, the mean deposition particle velocity must exceed a minimum (material-dependent) critical velocity to achieve deposition. This suggests that sufficient kinetic energy must be available to plastically deform the solid material and/or to disrupt the surface film. Below this critical velocity, incoming particles are only able to cause a deformation and abrasion of the substrate. If the critical velocity has been exceeded, Assadi et al. (2003) suggest that adiabatic shear instability at the interface could be the bonding mechanism for particles in GDS. It is proposed that bonding occurs through the formation of a jet at the interface. This is a required condition for successful bonding in shock-wave consolidation – a similar, intensive plastic deformation process. The jet that forms is a narrow, nearly planar or two-dimensional region of very large shearing known as an adiabatic shear band (ASB) (Wright 2002). Further, it is known that jet and vortex formation occurs at the interface under powder shock compaction (Meyers et al. 2001, Benson et al. 1997). Thus, the adiabatic shear band formation is considered to be the primary mechanism of shock compaction (Meyers et al. 2001). The formation of an adiabatic shear band is an unstable phenomenon in materials that deform at high rates. Here strain hardening is overcome by thermal softening (Li et al. 2003). The microstructure evolves in the following way during a high-strain rate deformation process. First, there are randomly distributed dislocations. Elongated dislocation cell formation then takes place, followed by elongated subgrain formation. Next, there is an initial break-up of elongated subgrains and finally a recrystallized microstructure. An adiabatic shear band is now formed. It is for this reason that the characterization of adiabatic shear bands is considered to be very important in bonding analysis.

In the GDS process, substrates are coated with powder mixtures consisting of particles with varying size and density. As shown earlier, the GDS particle consolidation processes are governed by the described mechanisms of ASB formation due to the shock consolidation of these particles upon impact. Therefore, the aim of the remainder of this chapter is to develop a shear compaction model for GDS of powder mixtures.

3.6 Impact Powder Compaction Model

The model of impact powder compaction is depicted in Fig. 3.11. Here smaller particles are seen to be bonded to neighboring particles and/or the substrate.

A larger spherical particle in the bimodal powder mixture then approaches and collides with this layer of smaller particles. This situation is typical of powder mixtures having differing particle sizes (see Fig. 3.11) as the smaller particles achieve greater velocities within the de Laval nozzle (Kreye 2004). The large particle acts as a hammer, causing the densification of the layer of smaller particles. If its velocity exceeds that of the critical velocity it goes on to bond with this layer and an adiabatic shear band is formed at the collision interface. The collision contact time can be approximated by Eq. (2.4). Van Steenkiste et al. (2003) and Kreye (2004) approximate the average impact stress by

$$P \approx \frac{\delta P}{A t_c} \tag{3.20}$$

where δp is the change in particle momentum as a result of the collision and is the cross-sectional area of the flattened particle. Particle flattening which occurs as a result of impact with the substrate is characterized by the ratio of the splat diameter to the equivalent particle diameter. This average particle aspect ratio has been found to range from 2.5 to 4 (Van Steenkiste et al. 1999). If a value of 4 is assumed, then the average impact stress is found to be

$$P \approx \frac{\rho v^2}{6} \tag{3.21}$$

where ρ is the density of the incident particle.

The shock-wave powder consolidation process is characterized by the velocity and strain rate upon impact. In a study by Gotoh and Yamashita (2001), the strain rates at the particle interface during particle impact are found to be in the range of 107–108 s^{-1} for a particle velocity between 500 m/s and 800 m/s.

The basic scheme of particle shock consolidation presented in Fig. 3.12 is similar to that of shock-wave consolidation (Tong et al. 2003), shown in Fig. 3.13. In this case, a flyer plate is mounted onto a nylon sabot and accelerated by a smokeless shotgun or pistol powder through a smooth bore barrel. Flyer plate speeds of up to 2 km/s can be obtained at the barrel exit (Tong et al. 2003). The densification of the powder layer is accomplished through the passage of a strong shock wave generated by the impact of the flyer onto the powder region.

As discussed by Schwarz et al. (1984), for a powder porosity greater than 0.1, a good approximation for the shock energy in the shock-wave consolidation process is given by

$$E_s = \frac{fP}{2\rho(1-f)} \tag{3.22}$$

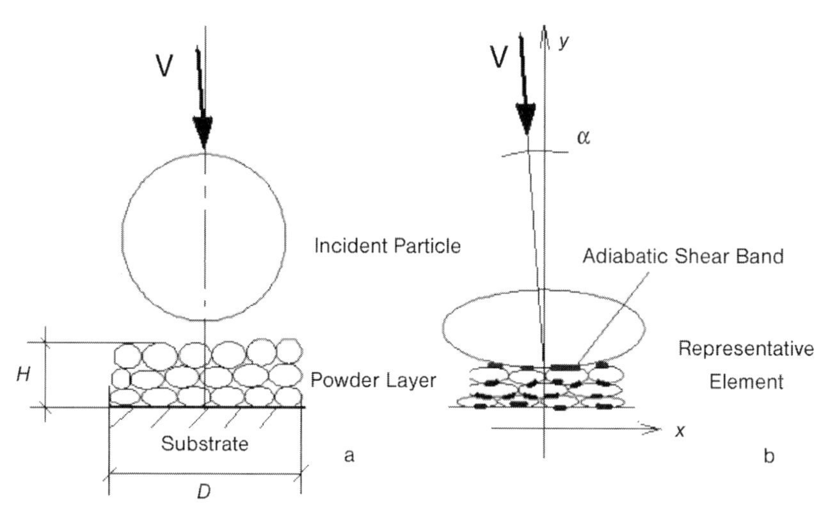

Fig. 3.11 A bimodal powder mixture GDS model. A large particle impinges upon the layer of smaller particles with the formation of adiabatic shear bands. (a) Before impact; (b) during impact particle consolidation. H and D are the height and diameter of a representative element, respectively, a is the incident angle, and V is the particle velocity.

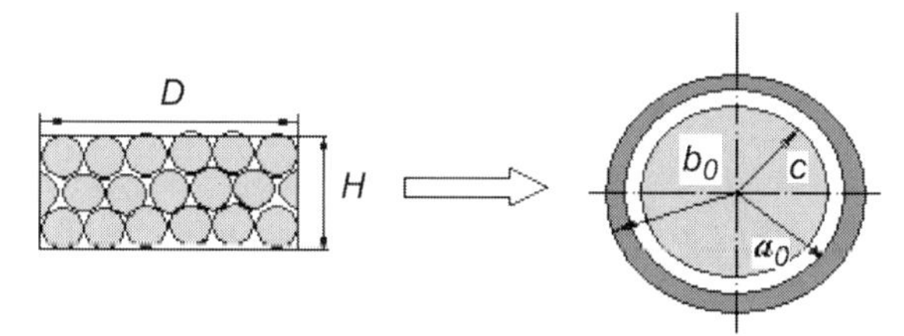

Fig. 3.12 Nesterenko (1995) sphere to calculate the microkinetic energy of the model for a bimodal mixture.

where f is the initial porosity of the powder layer, P is the compaction pressure of the shock wave, and ρ is the density of the powder material.

The majority of the shock-wave energy is dissipated as heat; some energy is also stored internally in microstructure defects. After a period of several hundred nanoseconds, the powder layer reaches a homogeneous temperature. Assuming adiabatic conditions, Tong et al. (2003) propose that the homogeneous temperature T can be evaluated at constant pressure P by

$$E_s = \int_{T_0}^{T} c_p(T)dT \tag{3.23}$$

where T_0 is the initial temperature of the powder layer and $c_p(T)$ is the specific heat of the powder layer. The relation describing the temperature dependence of the specific heat can be determined by curve-fitting experimental data (Weast 2000). Since the impact model given for the GDS process is similar to shock-wave consolidation, the above equations can be applied in the GDS case to determine the homogeneous temperature after the large particle collides with the powder layer. The compaction pressure of the shock wave is taken to be the average impact pressure in this case. The data used for calculation are given in Table 1.4, valid for temperature in the range of T_m 300 K, where T_m is the melting temperature of the metal. The details of calculation are shown by Maev and Leshchynsky (2006b) and Maeva et al. (2006).

3.7 Behavior of Consolidating Powder Under Compression

3.7.1 Constitutive Function

The powder densification and consolidation due to impact are the main processes of coating structure formation at the GDS. It is well established that in general a densification of the particulate media occurs in the following succession (Atkinson 1993): (i) rearrangement of grains (particles), (ii) fracture and rearrangement of grains, and (iii) deformation and fracture of grains. As the first stage of densification of a loose powder the particle rearrangement is obtained by an interparticle sliding which depends on the friction properties of the particles. The cohesion between the powder particles increases with the relative density. If the cohesion is very high, the powder system behaves like a sintered porous material (Seong-Jun Park et al. 1999). The deformation and fracture of particles are defined by their bulk properties. Thus, the main task of the powder densification analysis is to evaluate both the interparticle sliding (friction) and bulk properties of particles. In general, the compression of a powder is characterized by the compaction curve (pressure–density relationship). The behavior of powders under compression has been described on the basis of constitutive models of continuum mechanics (see, for example, Tszeng and Wu 1996). Many of these models were originally developed for the sintered powders. However, at the beginning of GDS powder consolidation the loose powder behaves differently than sintered ones. Many of the differences between a powder media and porous body originate from the individuality of the powder particles in the powder system. In addition to particle distortion (plastic deformation for metal powders), fragmentation (for ceramic and pharmaceutical powders), and/or crystallographic glide (for oxide aggregates), the collection of powder particles can move and rotate against each other, giving rise to interfacial friction or cohesion.

The continuum models determine the yield condition of porous medium using second deviator stress invariant J_2' and first stress invariant J_1. A general form of the yield function has been proposed in many works (see, for example, Tszeng and Wu 1996):

$$\phi = AJ_2' + B(J_1 + k)^2 - Y^2 = 0 \tag{3.24}$$

where A, B, and k are functions of the relative density ρ, and Y is the apparent flow stress, which is at least a function of the relative density. Also

$$J_1 = \sigma_{kk} = 3\sigma_m = 3p\ ,\ J_2' = \frac{1}{2}\sigma_{ij}'\sigma_{jj}'\ ,\ \sigma_{ij}' = \sigma_{ij}' - \sigma_m\delta_{ij} \tag{3.25}$$

where σ_m is the mean stress, σ_{ij} is the macroscopic stress, and δ_{ij} is the Kronecker delta function.

The various models concerning porous metals, as reviewed in Lee and Kim (1992), are all in forms similar to Eq. (3.24), except for the fact that k is set to 0. The physical reason for including the coefficient k in Eq. (3.24) is to reflect the limited interparticle strength when subjected to a tensile stress state. In order to use Eq. (3.24) for both porous materials and various powders, it is assumed that the apparent flow stress Y is a fraction of the flow stress for the individual particle, Y_0, i.e.,

$$Y^2 = \eta(\rho)Y_0^2 \tag{3.26}$$

One can see from (3.26) that the parameter $\eta(\rho)$ depends only on the relative density of the porous material and represents the square of the Y/Y_0 ratio. The specific forms of A, B, k, and η in Eq. (3.24) are greatly dependent upon the experimented and material data. The hydrostatic pressing of powder as well as the green strength of the powder compact has been utilized to determine the coefficients A, B, and k (Tszeng and Wu 1996). However, the above relations have been rigorously proved for loose powders.

In the case of GDS powder consolidation, the phenomenological model based on Eq. (3.24) may be used to describe the constitutive behavior of particle media during shear compression. The composition of the powder mixture is an important characteristic at the first stage of GDS consolidation during impact with the substrate where sliding and the local plastic deformation play an important role (Seong-Jun Park et al. 1999).

3.7.2
Yield Function and Property Estimations

To estimate the properties of GDS it is proposed that the yield function described by Eq. (3.24) be utilized with a new approach to determine the flow stress of individual particles, Y_0. Also, the primary mechanical characteristic

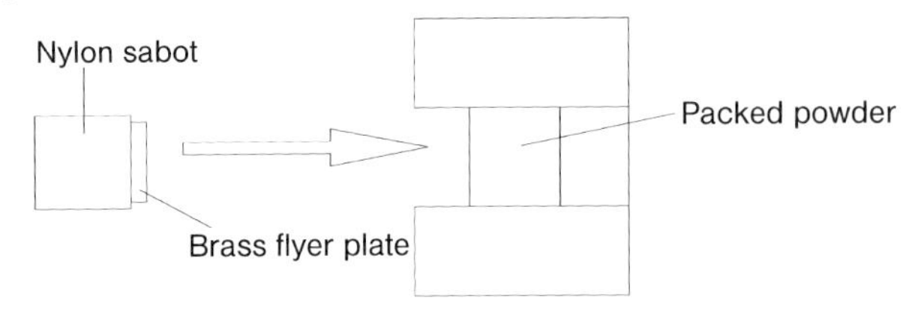

Fig. 3.13 The shock wave consolidation process.

of a powder body shall be considered to be its composition, k. First, we consider an isotropic (hydrostatic) compression. In this case the yield function (3.24) is transformed into

$$B(J_1 + k)^2 - Y^2 = 0 \tag{3.27}$$

According to Tszeng and Wu (1996), the function k is zero when the interparticle sliding during compression is stopped, resulting in the material becoming porous and consolidated. In the case, when the densification of this material is associated with a plastic deformation, the flow stress of individual particles, Y_0, can be calculated. The coefficients A, B, and η are assumed to be functions of the density:

$$A(\rho) = 2 + \rho^2 \quad B(\rho) = \frac{1-\rho^2}{3} \quad \eta(\rho) = \frac{\rho^2 - \rho_c^2}{1-\rho_c^2} \tag{3.28}$$

where ρ_c is the critical relative density and the yield stress of the powder is zero. Thereafter, Eqs. (3.26) and (3.27), along with the porosity functions (3.29), serve to fulfill the definition of the flow stress during an isotropic compression of the powder.

The shear powder compaction scheme is shown in Fig. 3.13. Analysis of shear compression made by Leshchynsky (1989) has presented an opportunity to define the apparent yield stress Y on the basis of Eq. (3.24) taking into account porosity ratios (3.28):

$$Y = \frac{\sigma_y \sqrt{3A\, B\,[3B + (B+3A)\cos^2\phi]}}{\cos\phi(4B + 3A)} \tag{3.29}$$

3.8
Consolidation Parameters of GDS and Shear Compression

The process of the shock consolidation is described by the scheme shown in Fig. 3.11. Representative element sizes of the powder media (Fig. 3.11) are given by $H = d_{\text{particle}} \cdot n$, where H is the height of the representative element,

$d_{particle}$ is the mean diameter of small particles ($\sim$ 10 μm), and n is the number of particle layers in one representative element ($n \approx 7$–10). If D is a diameter of the representative element (which is dependent on the size of the larger incident particle), the volume of the representative element with $D \approx 50$ μm is calculated to be $\delta_V = d_{particle} \cdot n \cdot \pi \cdot D^2/4 = 0.000196\ \text{mm}^3$.

The force created by the impact of an incident particle is $F = m \cdot a$, where m is the mass of the incident particle. An acceleration of this volume is calculated taking into account the duration of the impact ($\sim$ 50 ns, according Assadi et al. (2003)). Hence, the normal stress during the impact is defined as $\sigma_n = F/(\pi \cdot D^2/4)$.

In general, the normal stress σ_n is defined by the yield strength Y_0 of a material, which is in turn is a function of the particle temperature; indeed this factor must be accounted for. It is for this reason the ratio σ_n/Y_0 is used as a similarity parameter. In theory, in order to obtain forces similar in magnitude to that of GDS consolidation and powder compaction processes, it is necessary to achieve values of σ_n/Y_0 in the range of 10–40. This may indeed be possible through compacting powder by the shear compaction method (Fig. 3.14).

3.8.1 Estimation of Compaction Parameters

Calculation of GDS and compaction parameters was based on the following (Leshchynsky et al. 2005).

3.8.1.1 GDS Experiments

The shear velocity of a representative element of the powder volume may be represented by $V_x = -V \cdot \sin\alpha$ (see Fig. 3.10), where α is the incident angle of the impact particle. The shear displacement may be calculated as

$$\Delta = V_x \cdot dt \tag{3.30}$$

where dt is the duration of the impact ($\sim$ 50 ns). The shear strain $\gamma_{spraying}$ is

$$\gamma_{spraying} = \frac{\Delta}{H_i} = -\frac{V \cdot \sin\alpha \cdot dt}{H_i} \tag{3.31}$$

where H_i is the layer thickness within which a certain localization occurs. According to Wright (2002) and Gotoh and Yamashita (2001), reasonable values of H_i seem to be in the range of 2 μm. The results of this calculation are shown in Table 3.3.

3.8.1.2 Shear Compaction Modeling

The stress–strain state for a thin layer of powder media at shear compression is assumed to be uniform (Fig. 3.14). Therefore, the velocities during compaction are (at $y = h$)

$$V_n = V \cdot \cos\phi \qquad V_t = \sin\phi \tag{3.32}$$

The velocity distribution is

$$V_x = -\frac{V_c}{h} \cdot y = -V \cdot \sin\phi \cdot \frac{y}{h} \tag{3.33}$$

$$V_y = -\frac{V_n}{h} \cdot y = -V \cdot \cos\phi \cdot \frac{y}{h} \tag{3.34}$$

The strain rates (normal – ξ_i and shear – η_{ij}) may be determined as

$$\xi_x = \frac{dV_x}{dx} = 0 \quad \xi_y = \frac{dV_y}{dy} = -\frac{V \cdot \cos\phi}{h} \quad \xi_z = 0 \tag{3.35}$$

$$\eta_{xy} = \frac{dV_x}{dy} + \frac{dV_y}{dx} = -(V \cdot \sin\phi)/h \tag{3.36}$$

The shear strain in the case of shear compaction is calculated as

$$\gamma_{\text{compaction}} = \eta_{xy} \cdot t \tag{3.37}$$

where t is the duration of the compaction process, having a compaction velocity of $V_{\text{compaction}}$. Thus, the similarity between the cold spraying and shear compaction strain conditions is defined by

$$\gamma_{\text{spraying}} = \gamma_{\text{compaction}} \tag{3.38}$$

Hence, after inserting the above relations the shear compaction velocity is calculated to be

$$V_{\text{comp}} = -\frac{\gamma_{\text{spraying}} \cdot H}{\sin\phi \cdot t} \tag{3.39}$$

The results of calculations having compaction parameters $t = 10$ s, $H = 3$ mm, $\phi = 20°$, $40°$ and $50°$ are shown in Table 3.3.

3.9 Modeling Results and Discussion

3.9.1 ASB Width Evaluation

In the GDS process, many different powder mixtures may be successfully cold sprayed onto a substrate. The ASB widths may be calculated by employing the impact model for various bimodal powder mixtures. These computed values, given in Table 1.4, compare favorably with those in other severe plastic deformation processes (Wright 2002). Here the ASB width is calculated to be 7 μm. In another experiment (Levitas et al. 2000), various powder layers were densified using explosives. This process led to the formation of bands in the powder region that had approximate widths of 10–15 μm. In both cases, the calculated width of the bands was slightly higher than in three of the four situations presented in Table 3.2. These data allow one to estimate the area of adiabatic shear bands formation at the particle interface based on the volume fraction of the adiabatic shear bands Θ_{ASB}, as defined by Eq. (3.17) (Maev and Leshchynsky 2006c).

Tab. 3.2 Calculated ASB widths for different powders.

Incident particle	Powder layer	Incident particle diameter (μm)	Incident particle velocity (m/s)	Initial temperature of powder (K)	ASB width (μm)
Ti	Al	50	500	650	1.18
Ti	Cu	25	600	700	0.56
Cu	Al	20	700	675	12.54
Al	Cu	40	500	600	3.26

Tab. 3.3 Calculation of spraying and compaction parameters based on shear strain similarity.

Incident angle	Spraying (impact time ~ 50 ns)				Punch angle	Shear compaction			
α (°)	V_{spr} (m/s)	V_x (m/s)	Δ (μm)	γ	φ (°)	V_{comp} (mm/s)	H (mm)	t (s)	γ
1	600	10.4	0.52	0.26	10	0.45	3	10	0.26
2	600	20.9	1.05	0.52	40	0.24	3	10	0.52
3	600	31.2	1.56	0.78	50	0.31	3	10	0.78

3.9.2
Yield Stress of Powder Material

The main reason for modeling the powder media compaction that occurs during the spray deposition process is to determine how the density (porosity) and shear strain depend on the yield stress of powder material. To accomplish this, shear compression tests were performed. For all experiments the relative density ρ and porosity $\theta = 1 - \rho$ were calculated. Typical compaction curves are shown in Fig. 3.14. These curves have been found to be dependent on the porosity θ and the relative compaction normal stress σ_y/Y and may be

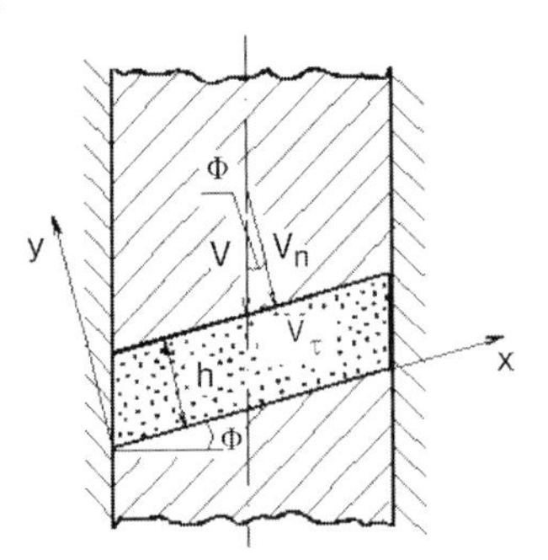

Fig. 3.14 Shear compaction method. The compaction of powder takes place in a container by punches with inclined faces.

shown to follow the power approximation:

$$\theta = M \cdot (\sigma_y / Y)^n \tag{3.40}$$

An increase in the shear strain (3.37) enables one to achieve higher densities with lower compaction stresses; this is due to the fact that the power coefficient n decreases. The specifics of this behavior, however, are strictly dependent on the composition of the powder mixture. The first evaluation of apparent flow stress Y – based on both Eqs. (3.24) and (3.28) and experimental results (Fig. 3.15) – show that interparticle sliding due to shear strains greatly diminishes the apparent flow stress of the powder media (Fig. 3.16) , as compared with those of solid copper. Al_2O_3 additives in the powder mixtures enhance this effect. The calculated yield stress Y_0 of a solid material for definite shear strains is shown in Fig. 3.17. Here the values of Y_0 are shown to be by two to three times lower then those of solid copper, 210 MPa. Also, Y_0 is dependent on porosity and shear strain (Fig. 3.18).

In accordance with the semiempirical equation (7.9) of Assadi et al. (2003), the critical velocity is dependent on the yield stress of the material and the apparent flow stress of the powder media that is consolidated during GDS impact. The investigations of Wu et al. (2006) and Lee et al. (2006) show the critical velocity to be greatly dependent on the yield stress. Thus, the comparison of GDS consolidation for one particle impact and powder ensemble impact reveals a sharp decrease in both the apparent flow stress Y and yield stress Y_0 and, consequently, in the critical velocity.

3.10 Concluding Remarks

The explosive nature of the structure formation processes in GDS seems to be the key to effective process analysis. From this viewpoint the energy approach

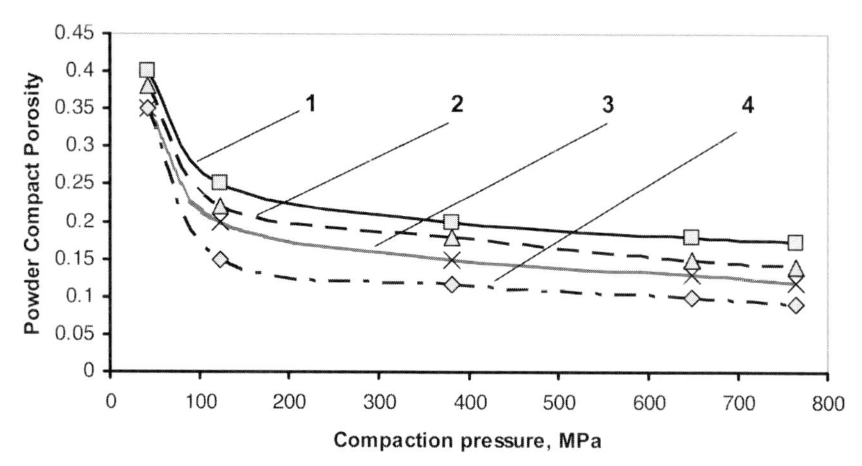

Fig. 3.15 Compaction curves for copper powder. (1) $\phi = 0^\circ$; (2) $\phi = 10^\circ$; (3) $\phi = 40^\circ$; (4) $\phi = 50^\circ$.

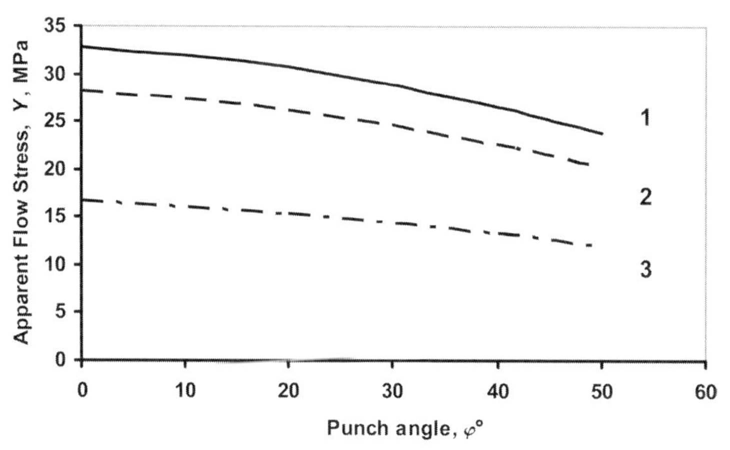

Fig. 3.16 Apparent flow stresses according to yield function of Tszeng and Wu (1996) (Eq. (1.50)) for the compaction of (1) Cu, (2) Cu+25%Al_2O_3, and (3) Cu+40%Al_2O_3 at the start of compaction process. Apparent density $\rho = 0.5$–0.6; compaction pressure $p = 30$–40 MPa.

of Meyers et al. (1999) is believed to be useful. It has been further shown that low-pressure GDS technologies exhibit similar shock consolidation features as high-pressure GDS. However, due to a lack of experimental results, reliable relationships between the structural properties of high and low-pressure GDS processes have yet to be developed.

Shear localization has been shown to be an important feature of gas dynamic spraying. More than this, it appears to be a necessary condition for successful bonding. In fact, the ASB width has been shown to vary between

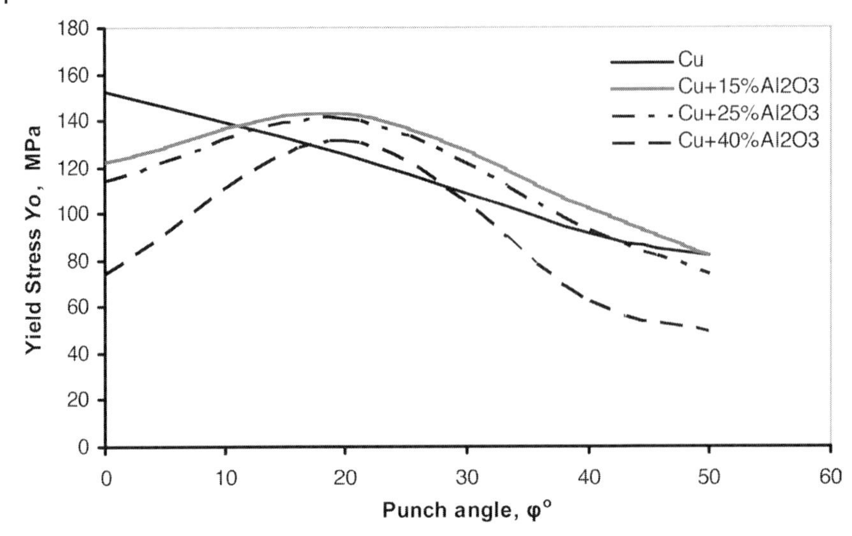

Fig. 3.17 Yield stress of solid phase Y_o according to Eq. (1.52) at the start of the compaction process. Apparent density φ_o = 0.5–0.6; compaction pressure p = 30–40 MPa.

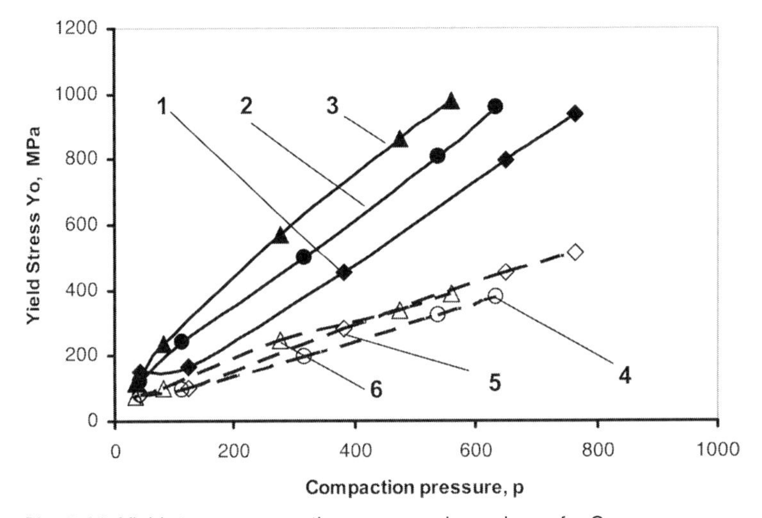

Fig. 3.18 Yield stress–compaction pressure dependence for Cu–Al_2O_3 powder mixtures during compaction. (1) Cu, $f = 0°$; (2) Cu+15% Al_2O_3, $f = 0°$; (3) Cu+25% Al_2O_3, $f = 0°$; (4) Cu, $f = 50°$; (5) Cu+15%Al_2O_3, $f = 50°$; and (6) Cu+25%Al_2O_3, $f = 50°$.

0.5 and 15 μm, revealing that great localization of particle deformation occurs in the GDS process.

Results of modeling reveal a decrease in the yield strength of powder media as compared to those of solid metal. This results in a lowered critical velocity parameter, ultimately influencing the shock compaction process during GDS.

4
Low-Pressure GDS System

4.1
State-of-the-Art Cold Spray Systems

Widespread adoption of GDS technologies suggests its potential for successful application in the industrial environment. Therefore, the development of a portable, low-pressure GDS system is of fundamental importance. It is important, then, to establish the relationship between the operating parameters of a portable GDS machine and the resultant microstructure of the formed spray coatings. To obtain high quality coatings, the configuration of a portable GDS system and the spray process parameters have to be carefully selected. Due to the sheer number of GDS process parameters, however, it is inevitable that a certain degree of trial and error is required to optimize the spray process for each specific coating–substrate combination. Presented below are some specific characteristics related to a portable, low-pressure system. These characteristics are subsequently compared with those of high-pressure GDS systems, a more established technology.

The stage related to the particle acceleration in a low-pressure GDS process is considered to consist of three subsequent processes:

1. gas and powder mixing;
2. acceleration of the particles within the divergent section of the spray nozzle;
3. particle movement within the free jet area (before and after contact with the substrate).

Each of these processes has its own characteristics and may be described by distinct parameters. Both the gas and particle velocities change during each phase of acceleration, as shown in Fig. 4.1. One can observe a substantial difference between the velocities of the gas and particle. This is due to the significant disparity in the drag forces that each encounters and must be taken into account when the parameters of the powder-laden jet are calculated.

There are two fundamental aspects that serve to distinguish the low-pressure spray process from that of high-pressure, or stationary, GDS:

Introduction to Low Pressure Gas Dynamic Spray Roman Gr. Maev and Volf Leshchynsky

ISBN: 978-3-527-40659-3

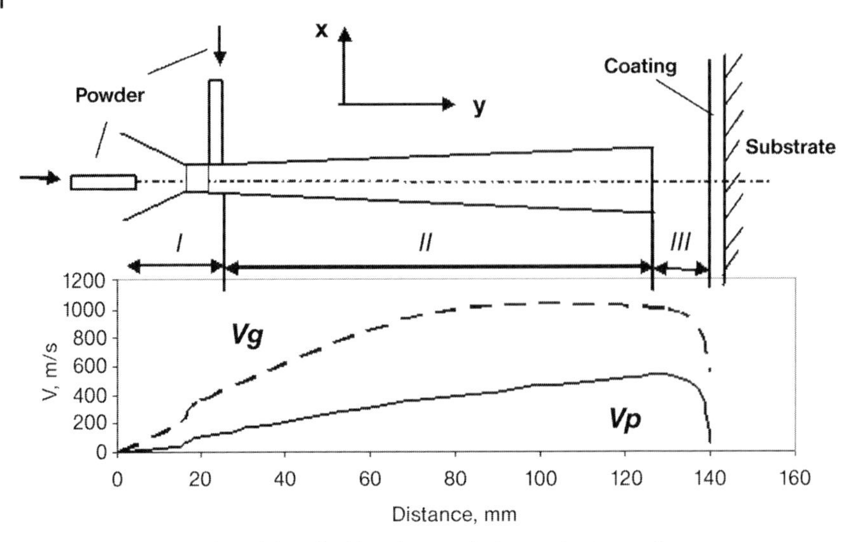

Fig. 4.1 The gas and particle velocities change during each phase of acceleration.

1. the utilization of low-pressure gas (0.5–1 MPa instead of 2.5–3 MPa);
2. radial injection of powder instead of axial injection (in most cases).

Through a close examination of these differences it can be noted that, due to a decrease in the gas mass flow rate, the particle content in the gas-particle jet is considerably higher in low-pressure systems. In general, the dependence of particle volume concentration on the powder mass flow rate (as shown in Fig. 4.2) reveals that the particle concentration is proportional to both the air flow and powder mass flow rates. In low-pressure spray processes both of these variables can be compared with those of traditional axial injected high-pressure GDS. In Fig. 4.3 the powder concentration and deposition efficiency of low-pressure GDS and of different variations of spray systems are presented (as developed by Alkhimov et al. (1994) and Van Steenkiste (2003)). The decrease in the gas flow rate due to the diminishing gas pressure can be observed to cause a sharp increase in the particle concentration of the gas-particle jet. It is known that the solid phase content in the gas-particle jet in traditional GDS is in the range of 10^{-5}–10^{-6} (by volume), while the content for low-pressure GDS systems is in the range of 10^{-4}–10^{-5}. For this reason the characteristics of gas-powder flow vary considerably. Hence, the particle impact parameters depend strongly on the type of GDS system that is employed.

As illustrated by Papyrin (2001), Van Steenkiste (2003), and Kay and Karthikeyan (2003), there are various features that characterize typical GDS systems. The main trait of the GDS system presented by Alkhimov et al. (1994) is that the gas and particles form a supersonic jet (with particle velocities ranging from 300 to 1200 m/s) whose temperature remains insufficient to

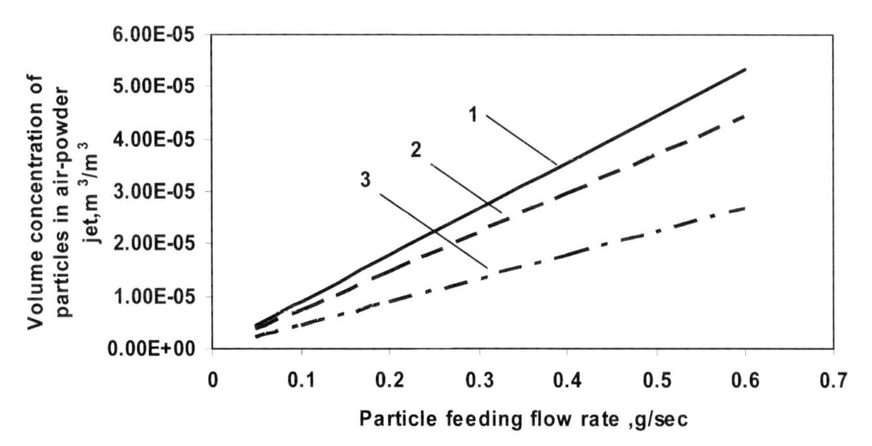

Fig. 4.2 The particle volume concentration in a powder-laden jet for low-pressure GDS. (1) Air flow rate 0.25 m^3/min; (2) air flow rate 0.3 m^3/min; (3) air flow rate 0.5 m^3/min.

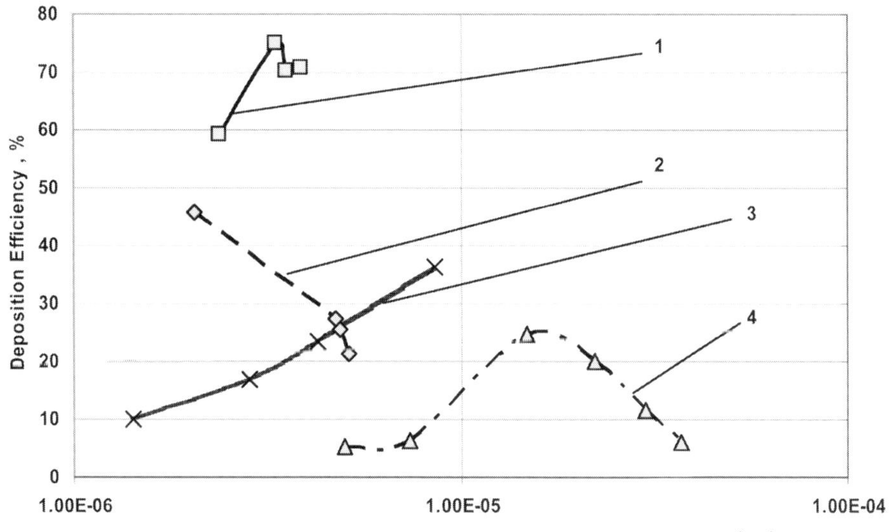

Fig. 4.3 Deposition efficiency of high-pressure GDS (1, 2, 3) and low-pressure GDS (4) processes. (1) Cu, particle size 63–106 μm, after Dykhuizen and Smith (1998), (2) Cu, particle size ¡45 μm, after Dykhuizen and Smith (1998), (3) Al, particle size 45 μm, after Papyrin (2001), (4) Al, air temperature 550°C, air pressure 7 bar, particle size 45 μm.

bring about thermal softening. Although the particles are introduced into the gas stream at subsonic velocities, the particle-gas jet is brought to a supersonic speed via the de Laval-type nozzle.

The initial design of the GDS system put forth by Alkhimov et al. (1994) was rather complex. However, Kay and Karthikeyan (2003, 2004) made significant improvements to this model by redesigning the nozzle. This design change allowed the heater mechanism to be removed from the main body of the gun. In this way, the spray gun applications proved to be considerably more flexible; this enabled its use within deposition geometries that would otherwise have been physically precluded in the previous version of the GDS system (Alkhimov et al. 1994). As a consequence, a bulkier and heavier electrical heater is required in order to heat the large volume of processing gas. Further, moving the gun and heater assembly requires a heavy-duty mechanical manipulator and, as a result, restricts the movement of the spray beam. Thus, the flexibility of the entire spray operation is highly restricted in this arrangement.

Finally, during the spraying of some materials such as aluminum, the powder particles may get deposited inside the nozzle on the walls, blocking the gas flow path. When this happens, the gas flow may be greatly reduced or even stopped, causing an abnormal increase in the temperature and pressure of both the heating element and the gun. Such a sudden increase in temperature and pressure can damage the gun and the heater, and also affect the safety of the operator.

The high-pressure GDS system developed by Van Steenkiste (2003) – or so-called kinetic spray system – is based on the Alkhimov et al. (1994) model. This system is able to produce spray coatings using particle sizes greater than 50 μm and up to about 106 μm. A further modified version of this device is capable of creating coatings using particles of larger size (up to 120 μm) and is designed to heat a high-pressure air flow up to about 650°C. After introduction of the particles, the heated air-particle jet is directed through a de Laval-type nozzle to achieve particle exit velocities between 300 m/s and 1000 m/s. The primary drawback of this system is that the spray material is injected into the heated gas stream prior to its passage through the de Laval nozzle. This brings about an inherent tendency for clogging, resulting in backpressure and spray gun malfunction. The use of large particles in the Van Steenkiste (2003) design may enhance this problem, requiring additional cleaning of the nozzle.

The second difficulty in CS devices is attributed to the low durability of the convergent throat portions of the nozzle. Because the heated main gas stream is under high pressure, injection of the powder itself requires a high-pressure powder delivery system. Unfortunately, these are quite expensive and are not conducive to the design of portable cold spraying devices. In the Van Steenkiste (2004) design, in an effort to overcome this shortcoming, a specially developed radial method is used for powder injection. This system involves injecting the particles by means of a positive pressure that exceeds that of the main gas pressure. Radial injection takes place within the diverging region of the nozzle, just after the throat. However, the solution of Van Steenkiste

(2004) does not eliminate the use of a high-pressure powder delivery system. Further, it introduces an additional difficulty – the necessity to permanently maintain a specific pressure difference between the positive injecting pressure and main gas pressure. Moreover, the value of this difference is strongly dependent on the location of injection point. Thus, in general, an additional adjustment of pressure is needed for different injection points, adding certain complications to the spray system.

The first portable, low-pressure CS system was developed by Kashirin et al. (2002) and referred to as a gas dynamic spray system. A GDS device comprises a compressed air source that is connected by a gas passage to a heating unit, and in turn fed into a supersonic nozzle. The divergent section of the nozzle is then connected by a pipe to a powder feeder just before the throat. When in operation, the compressed air is delivered to the heating unit where it is heated to approximately 600°C, and subsequently accelerated by the supersonic portion of the nozzle. The powder material is then transferred via the feeder to the supersonic nozzle where it mixes with the air flow and is accelerated toward the injection point on the nozzle. Within the nozzle the static pressure is maintained below atmosphere, ensuring that powder is effectively drawn in from the powder feeder. This pressure can be maintained only if the cross-sectional area of the supersonic nozzle in this portion exceeds that of the throat. The main difficulty associated with the system of Kashirin et al. (2002) is that the spatial uniformity of powder-air flow is quite low due to particle movement in the long powder passage. Both the particle-wall and interparticle friction result in particle accumulation in various places within the powder feeder conduit. The second difficulty is that precise control of the powder feeding rate is nearly impossible because particle movement within the powder feeder is strongly dependent on the length of the passage pipe, fluidity, and specific weight of powder or powder mixture. Additionally, the application of a heating unit where the outlet is connected to the convergent part of the nozzle does not facilitate intensive gas heating in a small coil heating unit.

Therefore, in order to improve the mechanism of gas heating, as well as the precision with which powder injection may be controlled, the low-pressure portable GDS system needs to be modified. Only after the mentioned shortcomings have been addressed may a truly practical portable gas dynamic spray system become a reality.

4.2 State-of-the-Art Powder Feeding Systems

One of the primary goals of low-pressure portable GDS systems is the development a powder feeding and metering device for use in delivering powder mixtures to the divergent part of the spray nozzle. A variety of vibratory powder feed systems have been known for many years. Such feeding systems, dating back the 1950s and 1960s, have included both rotationally vibrated bowls and linearly vibrated channels or troughs. In the field of GDS, a distinct need exists for similar portable feed systems.

Not only must this system be portable, it must be reasonably accommodating in that it should facilitate the use of numerous metal and ceramic powders in the spray process. Naturally, each of powder mixtures has its own particle size distribution, specific gravity, coefficient of friction, or other properties that affect its flow characteristics. Therefore, the feed system must ensure consistent particle flow even for powders having significantly different powder flow rates.

Typical designs of powder feed systems may be divided into two groups: (i) fluidized bed hoppers and (ii) dispensing hoppers.

A powder feeding system comprising an enclosed hopper and a carrier gas conduit was suggested by Fabel (1976). A schematic of this device is shown in Fig. 4.4(a). Here the carrier gas supply is connected to a conduit that extends through the lower point of a hopper and continues to the point of powder-carrier gas utilization. In order to regulate the amount of fluidizing gas that is supplied to the hopper, the pressure at a point in the carrier gas line is monitored, as per the method suggested by Dalley (1970). As the pressure within the gas line is dependent on the mass flow rate of solids, the change in pressure may be used to regulate its flow.

Although the feeding system designed by Fabel (1976) has good repeatability, various problems have recently surfaced, particularly with respect to the control and uniformity of the powder feed rate. One such problem is pulsation. This shortcoming, apparently due to a pressure oscillation, results in uneven coating layers. A modified system by Rotolico et al. (1983) (Fig. 4.4(b)) has eliminated this difficulty inserting an additional fluidizing powder near the conveying gas line.

The use of the fluidized bed hoppers in the high-pressure GDS systems has been proven effective in both academic laboratories and industrial environments alike. The results that have been obtained are summarized in a review by Karthikeyan (2005). However, despite this degree of success, dispensing hoppers with powder feeding rate controls have not yet been applied on a large scale.

The powder feeder designs of Schinella (1971) and Tapphorn and Gabel (2004) use a vibrating structure to move powder from the receiving surface

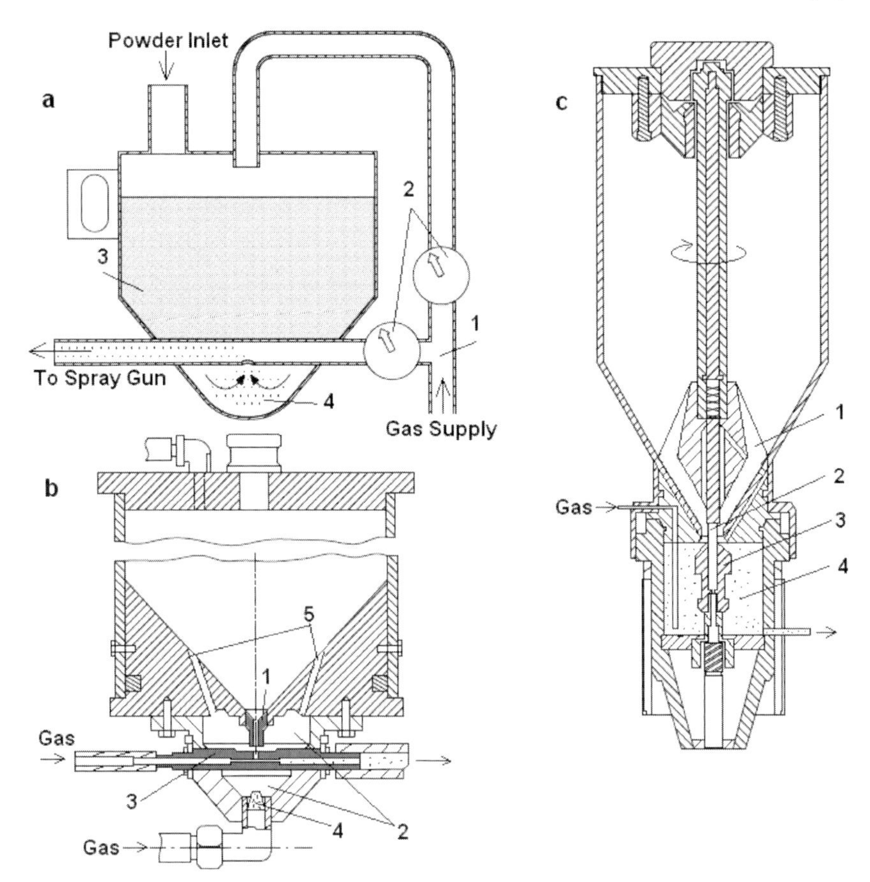

Fig. 4.4 Fluidizing powder feeding system design. (a) 3,976,332 – Fabel (1976): (1) carrier gas supply system; (2) pressure sensors and valves; (3) hopper; (4) fluidizing chamber. (b) 4,381,898 – Rotolico et al. (1983): (1) powder discharge fitting; (2) fluidizing chamber; (3) powder distribution system, (4) gas porous insert. (c) 3,501,097 – Dalley (1970): (1) powder feed passages; (2) tie rod; (3) powder chamber valve; (4) fluidizing chamber.

to the discharge channel (Fig. 4.5). Schinella (1971) describes the use of a hopper with an outlet channel and hemispherical cup for metering powder and powder flow control. The primary limitation of this powder feeder is that the consistency of powder metering is affected by the agglomeration of powder in the hemispherical cup. In some cases it may even become plugged.

An improvement in the design of the powder metering device has been made by Tapphorn and Gabel (2004). In this system, sieve plates are mounted within the hopper for precise metering of the powder as it moves into the vibrating bowl. The powder is metered through the sieve plates by a hopper vibrator that is in turn controlled by a level sensor mounted in the vibrat-

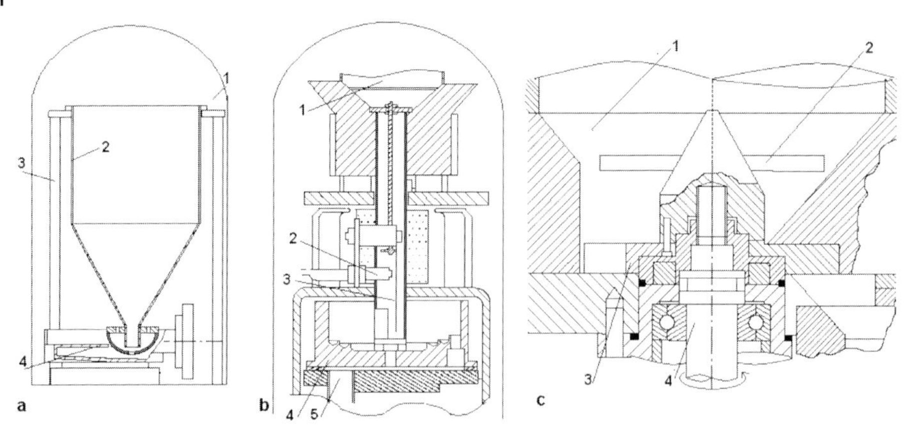

Fig. 4.5 Powder metering devices. (a) 3,618,828 – Schinella (1971): (1) pressurized chamber; (2) feed hopper; (3) hopper frame; (4) vibratory powder feeder-metering device with outwardly spiral path. (b) 6,715,640 – Tapphorn and Gabel (2004): (1) powder hopper; (2) powder feeding rate sensor; (3) flexible metal vane; (4) vibrating spiral bowl; (5) powder discharge outlet. (c) 4,808,042 – Muehlberger and de la Vega (1989): (1) powder hopper; (2) powder mixer; (3) powder feeder-metering disc; (4) powder feeder shaft.

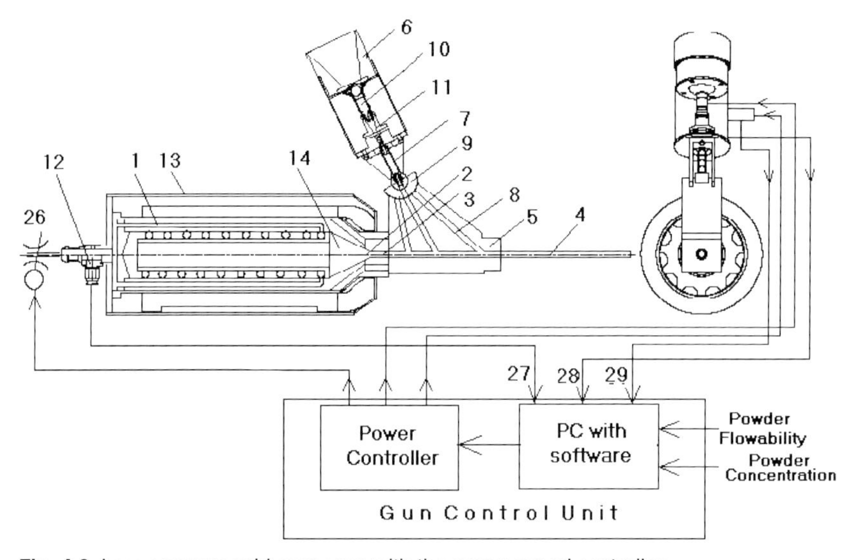

Fig. 4.6 Low-pressure cold spray gun with the programmed controller.

ing bowl. Unfortunately, the limitation of this design lies in its repeatability and accuracy. In fact, powder feeding rate measurement by a level sensor is strongly dependent on the agglomeration and fluidity of both powder and, in particular, powder mixtures. Moreover, in the case when the funnel is plugged with a flexible metal vane, the amount of powder flowing from the funnel tube to the receiving surface (and, in turn to the discharge channel) may not be adequately controlled.

4.3
Modification of the Low-Pressure Portable GDS System

A portable gas dynamic cold spray gun has been developed by the authors (Fig. 4.6). This system eliminates many of the inherent difficulties described above by minimizing the scatter of operating parameters and, thus, improving the spray gun efficiency. One advantage of this new system is that the powder flow rate is continuously measured. In this way, the powder flow rate and/or the flow rate of the pressurized gas can be adjusted in real time, leading to more controlled and efficient deposition by the spray gun. In this design a gas passageway extends through the spray gun. A gas port supplies pressurized air (or an alternate gas) to the passageway inlet. A nozzle in the passageway forms the pressurized air into a supersonic jet stream. The powder is fed at a controlled rate to the passageway, where it is entrained in the gas and exits the spray gun in the supersonic jet stream. The spray gun further includes a powder flow rate sensor that measures the powder flow rate of the powder. In the example of the spray gun shown in Fig. 4.6, the powder flow rate sensor includes a light emitter transmitting light across the duct through which the powder travels. A receiver mounted opposite the light emitter determines the flow rate of the powder based upon the measured light intensity. This signal is in turn processed by a specially developed controller (Fig. 4.6) which adjusts the gas flow rate and/or the powder flow rate to ensure that it is equal to that which is desired.

The portable GDS gun is connected to a pressurized gas source which supplies high-pressure air (or gas) to a heat chamber (1). The converging part (2) and the throat (3) of the nozzle are made of ceramic, while the diverging part (4) of the nozzle is made of steel. The diverging section of the nozzle extends through an outer housing (5) from which it is supplied with powder from a powder container (6). As the pressurized air or other gas passes through the nozzle, it reaches supersonic velocities and draws powder from the feeding hose (7) into the diverging part of the nozzle. The outer housing has multiple passages (8), each leading to the diverging part of the nozzle. Powder is selectively supplied to one of these multiple passages in the outer housing using a switch (9). The switch supplies the powder based upon the value of negative pressure at certain points of the air jet. The switch may be set manually, or automatically by means of a controller which is programmed to the expected negative pressure points along the diverging part of the nozzle. Depending upon the pressure of the gas source, the locations of negative pressure points along the diverging part of the nozzle may vary. The switch is set so that the selected passage coincides with the negative pressure point.

The powder from the container is fed to the switch through a vibrating bowl (10) and funnel (11) into the partial-vacuumed powder passages of the outer housing. The powder then mixes with the jet of conveyance air and flows through the nozzle where the diverging imparts supersonic velocities.

The jet of conveyance air from the pressurized air source is heated in the heat chamber. The compressed-air line contains a variable throttle (12) by which the flow impedance (e.g., the flow cross-section) is regulated from a controller according to a predefined value of either the volumetric flow of conveyance air and/or the volume concentration of the particles in the powder-laden jet. Oftentimes the controller is a computer having a processor, memory, and additional storage capabilities.

The heat chamber includes a serpentine or helical coil heating element mounted on a ceramic support and an insulation chamber which is located in an internal chamber housing (13). The air flows along the helical path defined by the heating element coil. The heated air exits the heater via the tapered chamber (14) which, together with ceramic insert (2), forms the convergent portion of the nozzle.

The powder supply system is shown in more detail in Fig. 4.7. The powder supply system includes a powder container (where the powder to be sprayed is stored in loose particulate form), a bowl vibration unit (15) for control of the powder flow rate, and a funnel (11) connected to a flexible hose (7). Additionally, two powder container vibration units (bowls) (16, 10) are incorporated into the powder supply system. Simultaneous control of the two vibration units provides precise and constant control of the powder feeding rate. Powder is fed into the lower vibration unit (10) to ensure that a certain level of powder is maintained by means of a sensor (17) which controls the operation of a main powder hopper (not shown). The rate of dispensing powder (powder flow rate) is additionally controlled by the removable lower bowl nose (18) with a variable diameter d of the hole and slot size a. The rate of dispensing powder is defined by fluidity of the powder.

The partial vacuum that exists in the partial-vacuum zone in the lower portion (19) of the pick-up chamber aspirates air from the atmosphere while being controlled by a flow throttle when passing into the partial-vacuum zone of the chamber. The chamber is fitted with a flow sensor (20) which generates a measurement signal corresponding to the rate of air flow from the atmosphere through the throttle (21) into the partial-vacuum zone of the chamber. Hence, the rate of powder passing through the powder passages (8) is controlled.

The pick-up device includes a powder metering unit (24) which detects the flow of powder particles in a measurement duct – this duct can be a simple glass powder transportation tube (23) connecting the funnel (11) to the

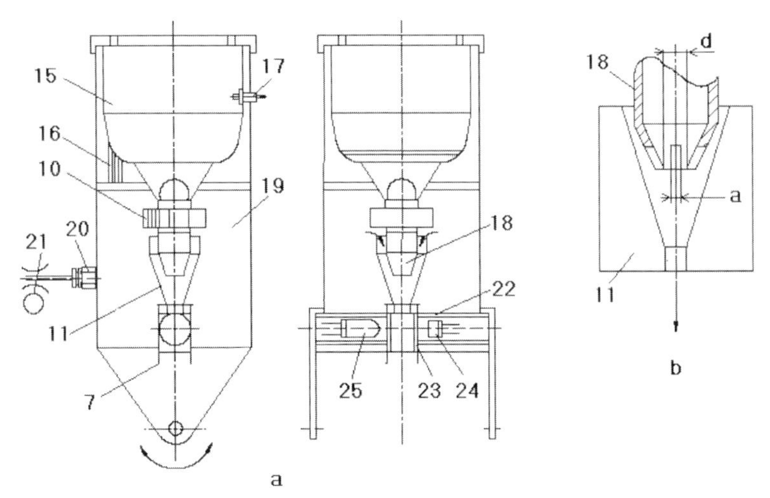

Fig. 4.7 Low-pressure cold spray gun. The powder supply system: (a) pick-up device and (b) removable nose and funnel.

powder aspirating hose (7) that is in turn attached to the powder switch of the outer housing. The powder-metering unit includes an infrared sensor (24) and an infrared emitter or light source (25) disposed within the channel. The infrared sensor can determine the mass flow of powder through the glass tube based upon the amount of light from the light source that is able to pass through the glass tube to the infrared sensor. Although an infrared light source and infrared sensor are preferred, other wavelengths of light or other waves could also be used. Optionally, an additional powder metering unit can be mounted in the pick-up housing on the opposite sides of the funnel (11). This additional powder metering unit, similar in nature to the previously described power meter, measures the powder dispensing rate ω_d from the lower vibrating bowl (10). The powder dispensing rate ω_d can then be compared to the conveyed powder rate ω_p. The amplitudes of the vibration units can be adjusted relative to one another in order to ensure that the powder dispensing rate ω_d is equal (over some short period of time) to the conveyed powder rate ω_p. This prevents clogging in the funnel.

The particle volume concentration significantly affects the deposition efficiency. The particle volume concentration in a powder-laden jet greatly influences the effectiveness of the GDS process. This is particularly true in the case of radial injection of powder by means of a conveyance air. Control of the concentration of particles is achieved through the regulation of both the rate of conveyed powder and the rate of conveyance air. The rate of the conveyed powder ω_p is substantially dependent on the powder dispensing rate ω_d and the rate of conveyance air. Also, the powder rate is approximately proportional to the rate of conveyance air in the partial-vacuum zone of the pick-up housing chamber. Therefore, the conveyance air must be adjusted to obtain the desired particle volume concentration in the powder-laden jet. The controller can automatically set the rate of conveyance air by means of the adjustment motor and throttle (26) (Fig. 4.6). On the other hand, the controller can set the powder dispensing rate ω_d by adjusting the amplitudes of the vibration units. In this way, the rate of conveyed powder ω_p may be maintained in permanent balance with that of the dispensing rate, i.e., $\omega_d = \omega_p$. The rate of conveyance air may be additionally regulated by a changing the location of the injection point via the outer housing switch. This may be done either manually or automatically.

Thus, the controller (Fig. 4.6) regulates the powder feeding flow rate, carrier air flow rate, and the powder conveyance air in the partial-vacuum zone of the pick-up chamber. This regulation is based on both measurement signals (27, 28, and 29) and preset specifications and is ultimately managed by means of the vibration units and two throttles. In actuality, the controller receives inputs regarding the desired powder dispensing flow rate ω_d (for instance in g/s), and the necessary volume concentration of the powder value C_v. On the other hand, this carrier air flow rate for the air passing through the powder/air duct (1) can be determined as

$$C_v = \frac{\omega_p}{\rho_p \, \omega_{\text{air}}} \tag{4.1}$$

where ω_p is the particle feeding flow rate from the funnel (Fig. 4.7(b)), ρ_p is the material density and w_{air} is the carrier air flow rate controlled by air pressure and throttle.

An alternative heat chamber design is shown in Fig. 4.8. The heat chamber includes the helical coil-heating element mounted on a ceramic tube within a carrier air transportation pipeline (see Fig. 4.8). The carrier air transportation pipeline is in turn mounted inside the internal chamber housing. The air flows in from the entry line between the internal chamber housing and the pipeline. The air then enters the forward end of the heat chamber pipeline and flows rearward within the helical heating coil. Once at the rearward end of the heat chamber pipeline, the air enters the ceramic tube and then travels in a forward direction through the ceramic tube, the tapered chamber, and finally, the ceramic insert which forms the converging part of the nozzle. Thus, the air gathers heat from the helical heating coil on three serpentine passes. This increase in the heating surface intensifies the heating of the air and consequently increases the temperature of carrier air up to 650–700°C in the portable heating chamber. Finally, the system also incorporates safety features for the protection of both the system and the operator. In the case of an abnormal increase in the temperature of the gas, the control system switches power supply off and renders a warning signal.

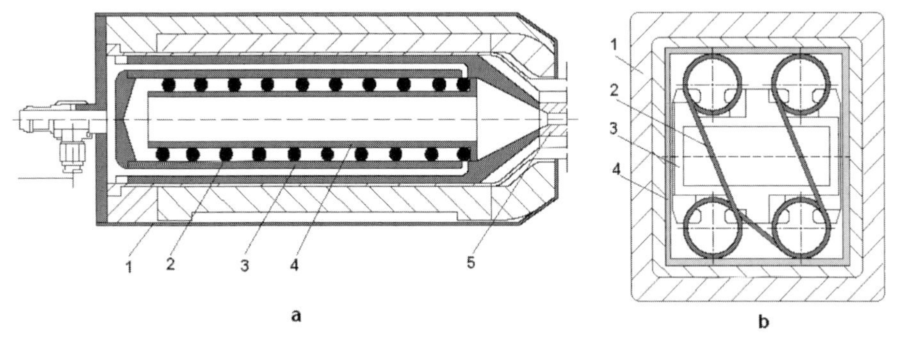

Fig. 4.8 Low-pressure cold spray gun. Heating chamber design. (a) Helical coil-heating design: (1) housing; (2) helical coil-heating element; (3) and (4) ceramic tubes; (5) nozzle insert. (b) Square chamber design: (1) multilayer insulation; (2) heating element; (3) heating element support; (4) ceramic housing.

4.4 An Industrial Low-Pressure Portable GDS System

The industrial low-pressure GDS system and its main components are shown in Fig. 4.9. This is reliable, rugged model of a portable GDS machine for industrial applications which was developed by Centerline Windsor Ltd. (Windsor, ON, Canada). The all-metal enclosure provides outstanding durability, ensuring that it can be moved from worksite to worksite quickly, easily, and without risk of damage. A touch screen display panel provides convenient and precise control while monitoring the machine's settings and diagnostics. Through the use of electronics, the digital controller ensures peak system performance by monitoring air pressure, air temperature, powder feed rates, etc. The SST spray gun incorporates an ergonomic and durable design which is intended to promote operator control and feel. With two mode selection buttons and an easy-to-operate trigger, it is comfortable for both right and left handled operators. The hopper and powder delivery system are equipped with programmable feed control and is sure to provide a long time of trouble-free performance. Dual 400 ml hoppers feed custom powder and abrasive compounds which are supplied in convenient hopper refills that can be easily replaced when material replenishment is required. An integrated "explosion-proof rated" system is used to provide quick and easy worksite clean-up. Also, in this system the collection and disposal of collected residual material may be done both safely and quickly. Finally, the touch screen control provides precise manipulation of the vacuum system.

a

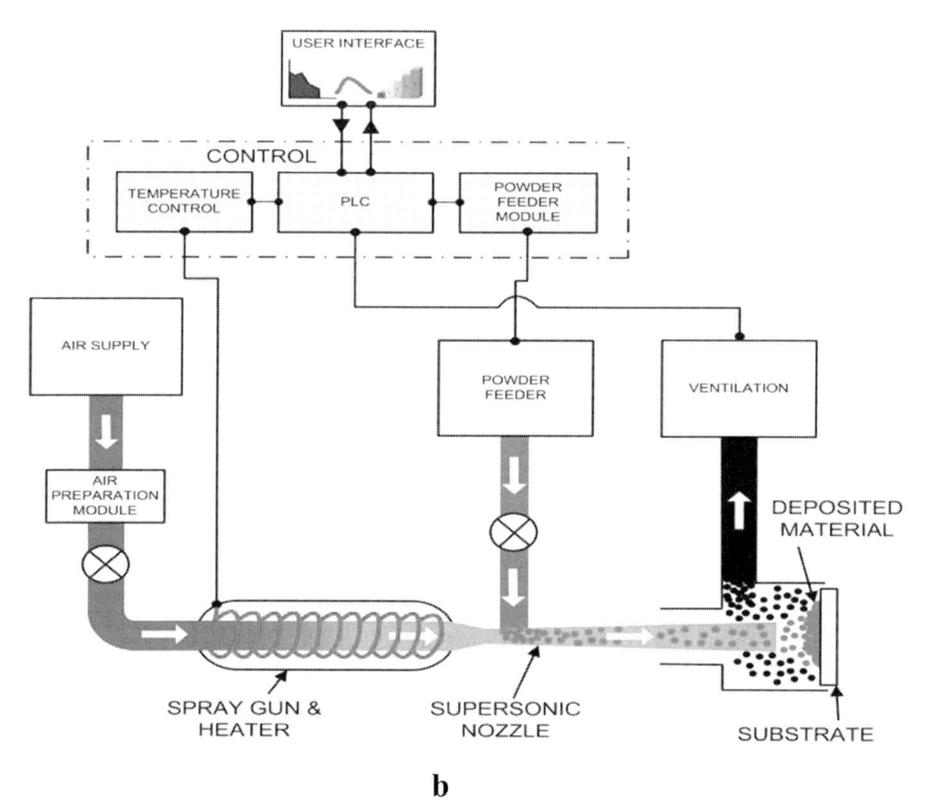

b

Fig. 4.9 Portable GDS SST machine (Centerline Windsor Ltd.). (a) General view; (b) principal scheme.

5
General Analysis of Low-Pressure GDS

5.1
Statement of Problem

It is a well known fact that coatings in GDS are formed due to significant plastic deformation in both the sprayed particles and the substrate, resulting from the impact of accelerated particles. In fact, these high impact pressures bring about direct contact between pure metals through the break-up and subsequent dispersion of a thin oxide film on both the particle and substrate surfaces, in turn leading to the creation of strong metallic bonds (Dykhuizen and Neiser 2003). On the other hand, the intensive deformation of the particle and substrate leads to increased strain and temperature at the localized contact areas. Several numerical simulation studies have indicated that the temperature at the contact zone, caused by the adiabatic shear during deposition, can reach values near the melting point of the sprayed materials, resulting in a thin melted layer (Grujicic et al. 2003, Alkhimov et al. 2000). Based on the metallic jetting observed with the impact of cold sprayed $TiAl_6V_4$ particles, Vlcek et al. (2003) suggested that this jetting may be attributed to melting at the impact interface. However, the occurrence of jetting may be a result of localized deformation under dynamic loading at the interface surface (Walter 1992). It should be noted, however, that strain localization may not be limited to the interface. Adiabatic shear bands may also occur in bulk particles (Walter 1992) as well as in powder granular media (Hu and Molinari 2004). The lack of strain localization analysis – particularly with respect to its influence on the structure of the formed coatings – limits our present understanding of the deposition mechanism in the GDS process.

Since the common approach to GDS coating formation analysis is related to the interaction between a single particle and the substrate, it is rather important to study adiabatic shear bonding processes in powder layers, taking into account the structure of the granular media, i.e., particle ensembles. In fact, if the air-particle jet consists of particles of various sizes, the nature of the particle interaction and coating formation depends on the behavior of particle ensembles upon impingement (Kochs et al. 1975). In the present approach we attempt to combine the theories of fluid mechanics that are relevant to solid

Introduction to Low Pressure Gas Dynamic Spray Roman Gr. Maev and Volf Leshchynsky

ISBN: 978-3-527-40659-3

particles suspended in a supersonic gas stream with the theory of particle consolidation during the spraying process.

The current GDS technology is of fundamental importance to create a very high speed gas that can accelerate the injected powders to velocities exceeding the critical velocity (about 500–800 m/s). This is normally achieved by expanding helium or nitrogen in a Laval nozzle. The pressure ratio for this expansion is in the order of 23. Due to its light weight, helium is a much easier gas with which to work. It is however very expensive, making the use of this process within an industrial environment economically impractical.

The critical gas velocity that must be achieved in GDS is material dependent, being affected by particle size and density, elastic–plastic properties, and powder composition (Grujicic et al. 2003, Stoltenhoff et al. 2002, Wu et al. 2006). In fact, by using relatively large particles in GDS processes using air, the minimum kinetic energy necessary for deformation and bonding with low particle critical velocities may be easily obtained (air GDS processes were first referred to as kinetic spraying by Van Steenkiste et al. (1999)). For critical velocities with Mach numbers 1.2–2, the air-pressure ratio is calculated to be in the range of 2.4–7.8. Thus, in order to achieve critical particle velocities using an air stream it is necessary to use a GDS system with air pressure parameters of about 5–8 bars.

One of the primary advantages of radial powder injection in low-pressure GDS (LPGDS) technologies is the high durability of the nozzles and simple design of the powder hoppers (Maev et al. 2004). Also, low air pressure and radial injection of the powder leads to decreased gas consumption, sharply improving the economical feasibility of LPGDS processes within industrial environments.

Thus, the purpose of low-pressure GDS analysis is to bring about a more thorough understanding of the effect the primary parameters have on structure formation when using various powder mixtures. These parameters are believed to be gas temperature, particle velocity, particle concentration in the powder-laden jet, and powder composition.

5.2 Experimental Procedure

The spraying of Al-based powder mixtures on steel and aluminum substrates was performed by Kashirin et al. (2002). Here an apparatus was utilized which incorporated a DYMET gun to inject powder into the divergent part of a supersonic nozzle. The experimental setup and schematic of the spray process are shown in Figs. 5.1 and 5.2.

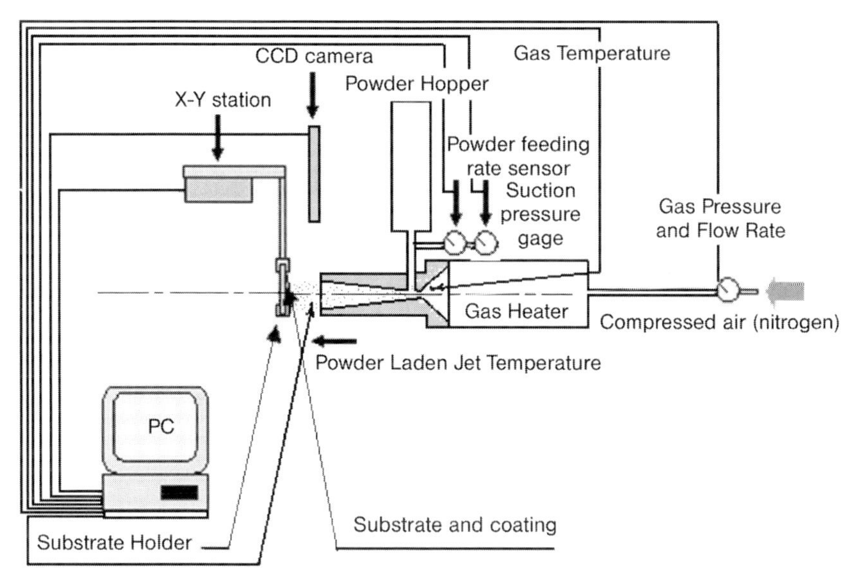

Fig. 5.1 Experimental setup scheme.

The powder mixtures were supplied by a powder hopper and injected into the supersonic portion of the nozzle near the throat area by means of the negative pressure that resulted from the stream of compressed air passing through the nozzle. Once introduced, the drag effect acts to accelerate the injected particles within the high velocity air stream. To increase the air velocity and, ultimately the particle velocity, the compressed air was preheated to temperatures ranging from 100°C to 700°C. The pressure and the temperature of the compressed air were monitored by a pressure gauge and a thermocouple positioned inside of the gun. The gun was installed on an X–Y manipulator to drive the air-powder jet over the substrate surface (Fig. 5.1). The compressed air pressure was kept constant at 0.5 MPa. It is important to note that the air pressure was lower than that in other gas dynamic spray systems (usually in the range of 2–5 MPa according to Papyrin (2001)). This feature makes GDS with a DYMET gun much more suitable for the industrial environment because a special high-pressure pump is no longer needed. Also, the air consumption is lower as well (about 0.2–0.3 m^3/min) when compared to spray systems using an axial injection of powder (1.3 m^3/min; Vlcek et al. 2003). The particle velocities at the exit of the supersonic nozzle were in the range of 550 m/s, as measured with a laser doppler velocimeter. The powder feeding rate varied in the range of 0.5–1.5 g/s, while the standoff distance from the exit of the nozzle to the substrate was held constant at 10 mm. A rectangular nozzle with an exit aperture of 3.5 mm × 10.0 mm was used.

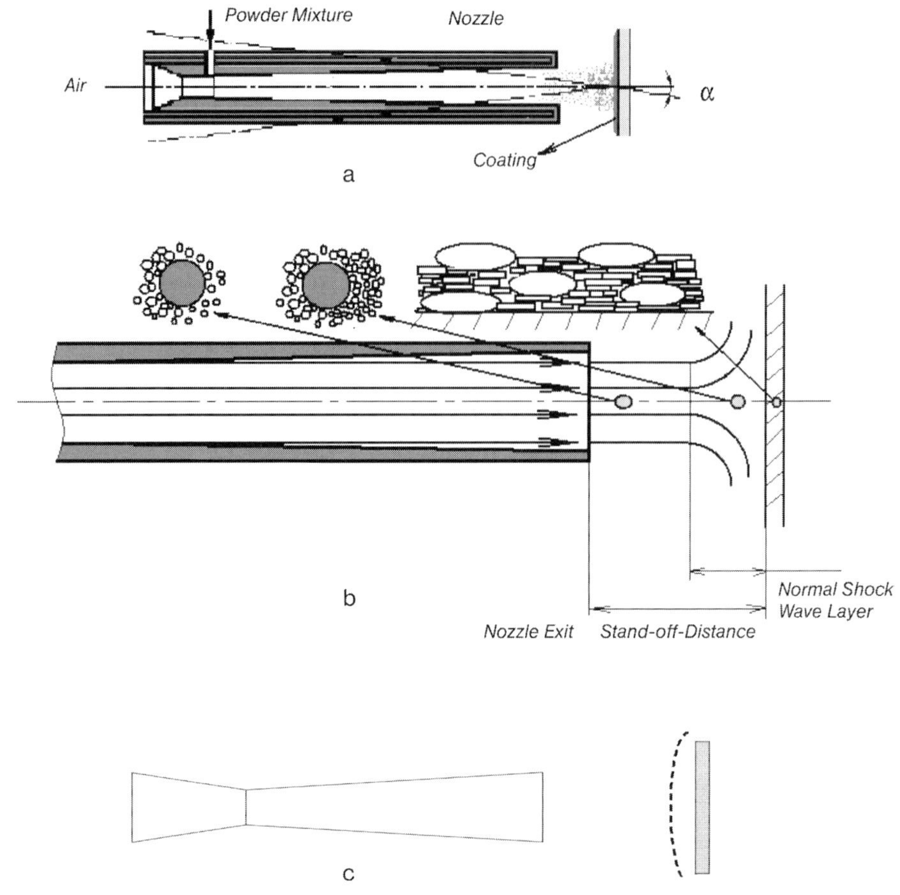

Fig. 5.2 Composite coating structure formation model for spraying powder mixture. (a) General scheme of powder jet and coating. (b) Particle arrangement in powder jet; nozzle-shock wave – substrate geometry.

Commercially available aluminum powder (−45 μm), zinc powder (5–10 μm), and tungsten powder (3–10 μm) were used as base materials. The Al powder used for coating deposition was sieved to sizes ranging from 40 to 60 μm to limit the effects of particle size distribution. SEM pictures of the powders used are shown in Fig. 5.3. Mild steel and Al were used as substrates. Prior to spraying, the substrate surface was sandblasted using 300 mesh alumina grits. Two-component Al–Zn and three-component (W–Al–Zn and Ti–Al–Zn) powder mixtures were used in the experiments. The deposition efficiency of GDS was measured on grit blasted 3-mm-thick aluminum substrates. The substrates were cleaned with acetone and methanol, dried in

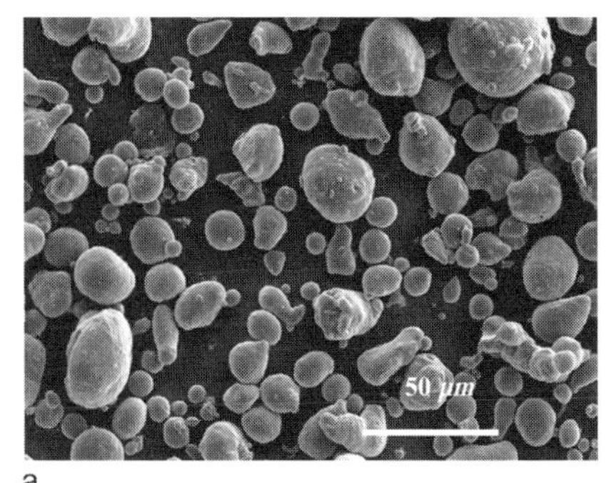

a

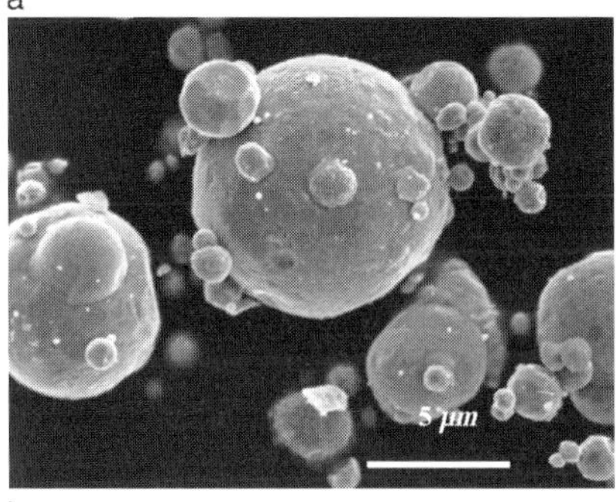

b

Fig. 5.3 SEM images of base powders for powder mixtures. (a) Al powder (×500); (b) Zn powder (×5000).

hot air and weighed before and after processing. A 25 g portion of the powder mixture was measured to be used for each sample.

Characterization of the coatings was based on microhardness measurements. The coating structure was examined using a light microscope after polishing and 10 s etching with Tucker's reagent. The phases of the coating were identified by X-ray diffraction.

5.3 Experimental Results

The GDS experimental results were analyzed with respect to the effects of air temperature T_g, powder jet composition, powder jet incidence angle α, and stand-off distance x_d. The air temperature influences the air and particle velocity, which are to be as high as possible. However, optimization of particle interaction and bonding during the spraying process may be achieved through the intensification of particle deformation and interparticle sliding. These processes are controlled through the composition of the powder impinging the

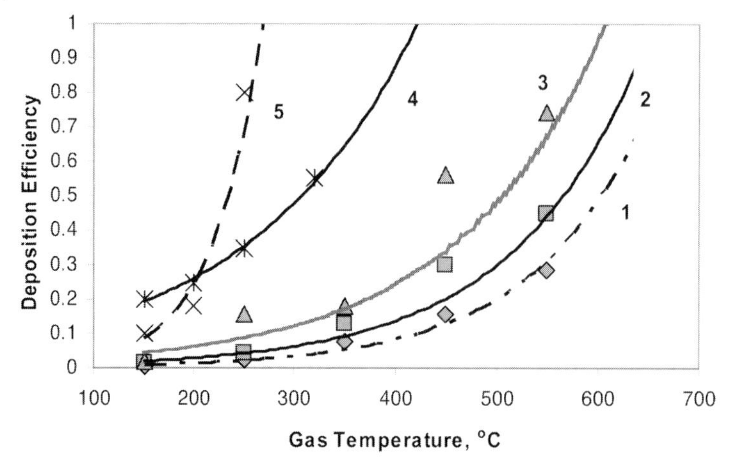

Fig. 5.4 Deposition efficiency for powder mixtures: (1) powder mixture with weight concentration $0.25Al_2O_3 + 0.5Al + 0.25Zn$; (2) powder mixture with weight concentration 0.5Ti + 0.25Al + 0.25Zn; (3) powder mixture with weight concentration 0.5W + 0.25Al + 0.25Zn; (4) Cu, after Borchers et al. (2003); (5) Ti, after Li and Li (2003).

substrate, particle sizes, and the incidence angle. In an effort to better understand the process, the following dependences are to be examined:

i. Deposition efficiency versus air temperature (Fig. 5.4) and powder feeding rate (Fig. 5.5);
ii. average single pass thickness of deposited layer versus air temperature (Fig. 5.6);
iii. hardness versus distance from the surface (Fig. 5.8).

5.3.1 Deposition Efficiency

Experimental results demonstrate various features common to powder mixtures sprayed at low air pressures: the deposition efficiency χ_D of GDS is found to be low for temperatures ranging from 100 to 500°C. However, a sharp increase can be observed at temperatures higher than 600°C, becoming comparable with the values of χ_D for Ti and Cu (Stoltenhoff et al. 2002). It is important to note that nitrogen was applied for GDS of Ti (Li et al. 2003) and copper (Borchers et al. 2003). It is known that the nitrogen and helium propellant gases allow higher gas and particle velocities to be achieved, as compared with that of air (Dykhuizen and Smith 1998). In fact, particle velocities reach critical values at temperature of 300°C for nitrogen. The deposition efficiency of Ti is above 0.6 at this temperature (Fig. 5.4) (Li et al. 2003). For air GDS at a low pressure of 0.5 MPa, a χ_D value of ~0.6 is realized at air temperatures

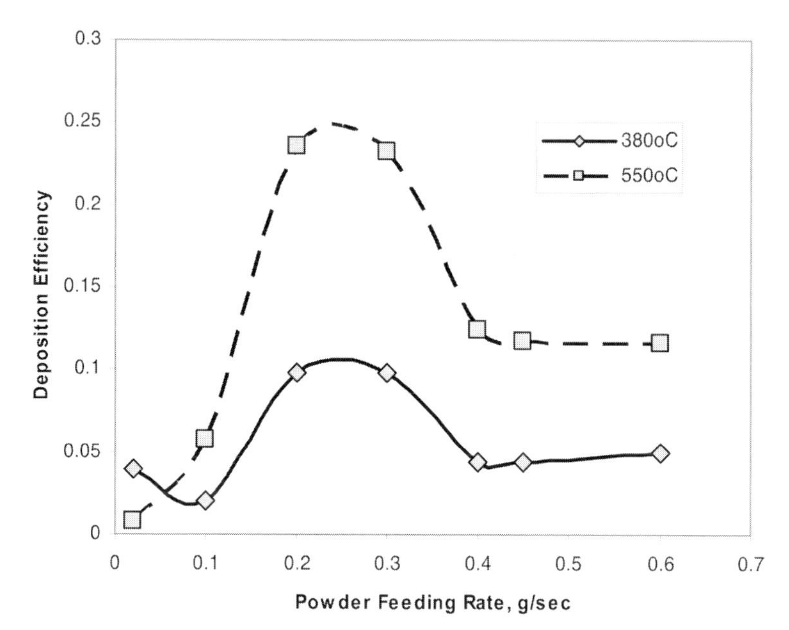

Fig. 5.5 Deposition efficiency versus powder feeding rate. Experimental data for Al-based powder mixture with weight concentration $0.25Al_2O_3 + 0.5Al + 0.25Zn$.

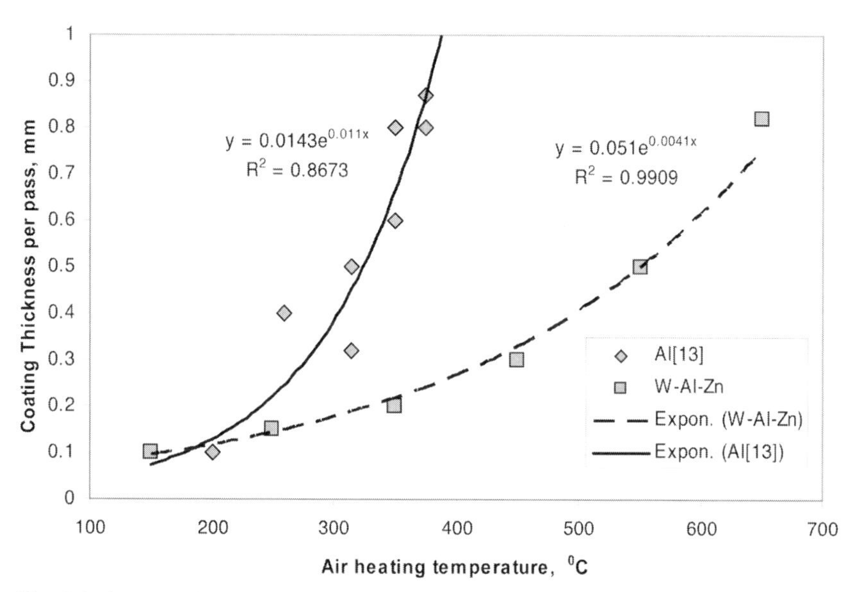

Fig. 5.6 Average single pass thickness – air temperature dependence. Experimental data and exponent approximations for Al (Van Steenkiste et al. 1999) and Al-based powder mixture with weight concentration 0.5W + 0.25Al + 0.25Zn.

of 650°C. Thus, we can assume that the critical velocity threshold of 500 m/s (Dykhuizen and Smith 1998) is achieved at this temperature.

The dependence of the deposition efficiency on temperature is accurately described by an exponential approximation for all powders and powder mixtures (Fig. 5.4). Simulation veracity is in the range of R^2 = 0.94–0.99. The deposition efficiency equation is approximated by

$$\chi_D = \chi_{D_0} e^{-Q_D/RT_0} \tag{5.1}$$

where χ_{D_0} is the baseline deposition efficiency. Q is an activation constant of the GDS process and is fixed for definite powder mixtures.

Approximations of the experimental data for studied powder mixtures are shown in Table 5.1 for three types of powder mixtures.

Tab. 5.1 Parameters of the exponential approximation equation (5.1).

Number	Powder composition by weight	Baseline deposition efficiency, Eq. (2.1) χ_{D_0}	Activation constant of Eq. (2.1) Q_D (kJ/mol)	Exponent coefficient of Eq. (2.2) t_0	Activation constant of Eq. (2.2) Q_t (kJ/mol)
1	25%Al_2O_3+50%Al +25%Zn	21.95	29.07	3.19	12.9
2	50%Ti+ 25%Al+25%Zn	20.35	26.05	4.97	12.92
3	50%W+25%Al+25%Zn	21.62	23.4	6.93	12.83
4	Copper (nitrogen)	6,48	12.45	–	–
5	Titanium (nitrogen)	4513	37.5	–	–
6	Aluminum (air)	–	–	164.2	28.47

Analysis of the constants in Eq. (5.1) shows that the values of χ_{D_0} do not vary for powder mixtures 1–3, while the values of Q_D are slightly different. This means that critical particle velocity, defined by the gas temperature, is the same for all powder mixtures studied because they consist of Al and Zn particles which are deformed during impact with substrate during coating formation. The values of the activation constants reveal the kinetic parameters of the powder build-up process, which depend on both the type and volume concentration of the larger and heavier particles. Thus, the deposition efficiency of the powder mixture 3 (see Table 5.1) is high due to the higher kinetic energy of the heavy tungsten particles, resulting in the conversion of plastic work into heat and stored energy within the formed GDS coating.

5.3.2 The Effect of the Particle Mass Flow Rate

The effect of the particle mass flow rate, M_s, on the deposition efficiency is shown in Fig. 5.5. The maximization of the function $\chi_D = \chi_D(M_s)$ reveals that there is an optimal value of the particle mass flow rate, which provides an optimal concentration of particles in the powder jet. This is characterized by the particle-to-gas mass flow ratio, $\mu = M_s/M$, where M is the gas mass flow rate. If the air temperature in the nozzle outlet area is about 200°C (Van Steenkiste et al. 1999) and the air density is $\rho = 0.746\ \text{kg/m}^3$, then the particle-to-gas mass flow ratio is within the range of μ = 0.035–0.3 – this value is dependent on the particle mass flow rate M_S. The optimal value of μ for which the deposition efficiency is maximal is approximately 0.27. Van Steenkiste et al. (1999), on the other hand, showed that the particle-to-gas mass flow ratio for a kinetic spray process is about 0.06. Thus, it can be assumed that these values of the particle-to-gas mass flow ratio govern the mechanism of individual particle bonding during the coating structure formation process.

In spite of lower powder feeding rates in processes using a DYMET gun (as compared to usual GDS process with axial powder feeding), higher solid phase concentrations can be achieved in the powder-laden jet by approximately five times. It is suggested by the authors that the specific shape of the function $\chi_D = \chi_D(M_s)$ is a result of higher powder particle concentration at the normal shock wave layer. The particles rebound from the substrate surface and collide with particles from the powder jet, sharply decreasing their kinetic energy. As a result, these particles have reduced kinetic energies and are no longer able to consolidate and form a coating. It should be noted that in spite of the fact that low air temperatures correspond with low particle velocities, the χ_D maximum can be observed on all air temperature curves (Fig. 5.5). This means that even for low air temperatures some particles are able to reach the velocity threshold that is necessary for bonding. Thus, in this case it is the mechanism of consolidation between particle ensembles and the substrate which defines the coating structure formation process. Unfortunately, these mechanisms are not fully understood.

5.3.3 The Build-up Parameter

The average single pass thickness of a GDS-coated layer or build-up parameter t is one of the main parameters characterizing the capability for GDS build-up. Its dependence on air temperature is shown in Fig. 5.6. The semiempirical relationship which gives a good approximation of the experimental data is in fact found to be similar to Eq. (5.1):

$$t = t_o e^{-Q_t/RT} \tag{5.2}$$

The parameters t and Q_t of Eq. (5.2) are shown in Table 5.1. It can be seen from these data that the activation constant Q_t does not depend on the air temperature and, consequently, on particle velocity. At the same time the parameter t increases by about two times showing the influence of the powder composition on the build-up process. The comparison of the activation constants in Eqs. (5.2) reveals the same order of magnitude. These expressions, describing the deposition efficiency and average single pass thickness parameters by means of an exponential function, correctly reflect the behavior of certain physical parameters. For deeper understanding of the physical mechanism behind GDS it is necessary to develop new phenomenological models that include the interaction of particles during the deposition of powder mixtures.

5.3.4
Structure and Properties

The examination of the properties and microstructure of coatings deposited by GDS reveals some important features of composite layer formation (Fig. 5.7). For example, consider the structure of a two-component coating consisting of 45% Al and 55% Zn; the large Al particles are surrounded by small Zn particles. Through careful examination it can be observed that only about 5–10% of the Zn particles are located between the Al particles. Further, the concentration of Zn in the composite coating is about 30–40%. However, the collision frequency of the small Zn particles is much higher than that of the large Al particles due to a higher concentration in the powder jet. These data indicate that the Zn particles rebound more intensely than the Al particles during the powder jet–substrate interaction. This conclusion can be verified through careful microstructure examination of Al–Zn powder mixtures having different concentrations. In fact, the Al and Zn particles exhibit different rebound properties depending on both particle velocity and the air temperature. Thus, the final content of Al and Zn phases in coatings does not necessarily correspond with the initial composition in the powder jet – in fact, in this case a 15% loss of Zn content was observed. These data demonstrate the great importance of the rebound process that occurs during interactions between the powder-laden jet and the substrate.

The main principle governing the formation of GDS coatings is related to the extensive plastic deformation at the particle–substrate interface (Sakaki and Shimizi 2001). As stated by numerous authors (see the review by Assadi et al. (2003)), the bonding mechanisms in GDS can be similar to shock compaction. For this reason it is beneficial to choose powder compositions having optimized consolidation properties. The construction of such a powder mixtures is reliant on

i. the optimization of shock consolidation properties through the inclusion of large, heavy particles in the powder jet;

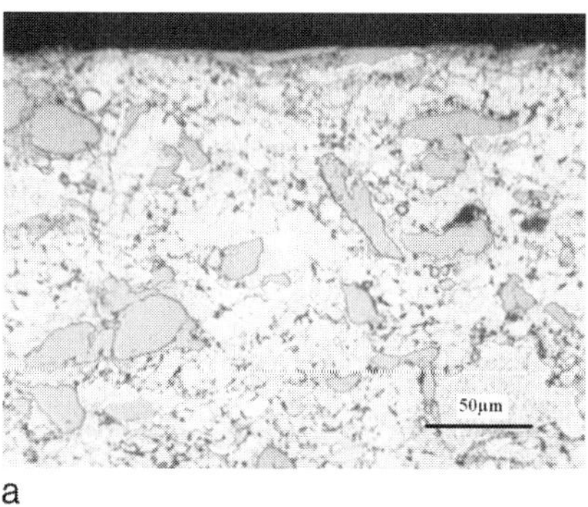

a

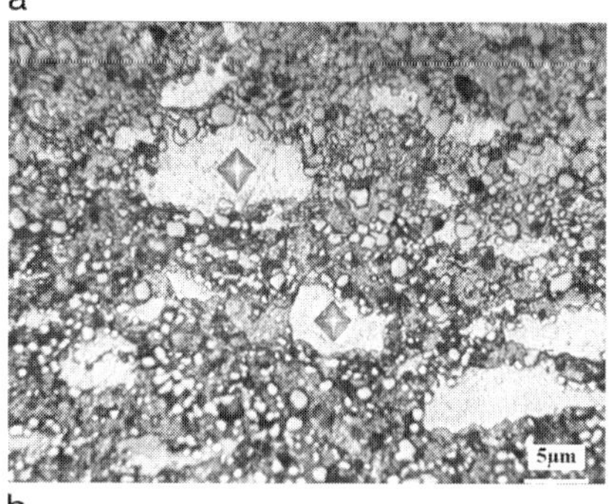

b

Fig. 5.7 Structure of composite coatings. (a) Al–Ti–Zn composite coating; (b) Al–W–Zn coating (not etched).

ii. choosing particles with high density that are able to adequately penetrate the powder mixtures; and
iii. solid phase reactions that influence the proper bonding of particles.

For this reason, we examine the microstructure of two-component Al–Zn coatings and three-component Ti–Al–Zn (Fig. 5.7(a)) and W–Al–Zn (Fig. 5.7(b)) coatings. The processes of structure formation result in different hardness levels of the deposited coatings (Fig. 5.8). The distribution of hardness is relatively uniform for pure aluminum coatings (Van Steenkiste et al. 2002) and composite W–Al–Zn coatings (the scatter is of 5–7 units). On the other hand, the hardness of Al–Ti–Zn coatings varies in the range of 70–120 $HV_{0.1}$. This is probably a result of mechanical alloying reactions, which may occur during the collision of Al, Zn, and W particles. Additionally, all coatings have a lower hardness in the vicinity of the substrate surface. This phenomenon is believed to be explained by the effect of the localization due to the severe particle deformation during the deposition process. At the same time, the hardness of the substrate is very high due to the surface strengthening that takes place during the grit blasting process. Phase microhardness examination results reveal the effect of W and Ti on the microhardness of the aluminum phase in the range of

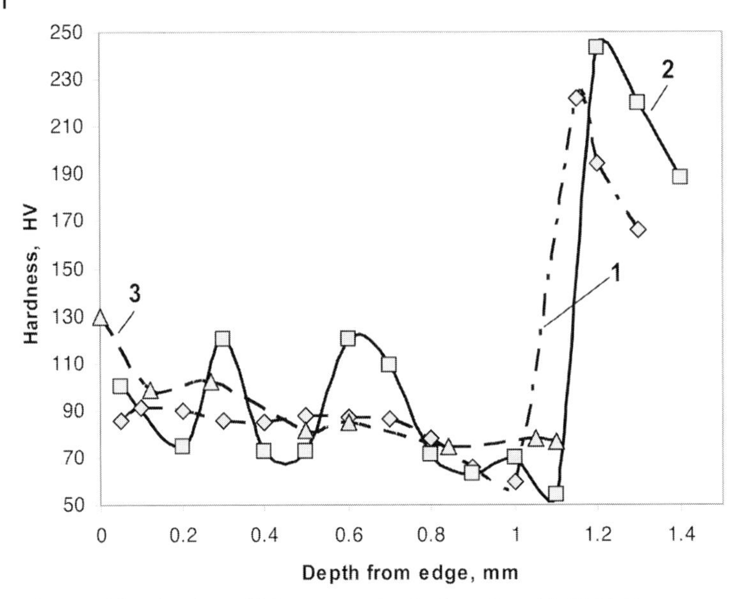

Fig. 5.8 Microhardness $HV_{0.1}$ versus depth of coating. (1) Al–Ti–Zn coating; (2) Al–W–Zn coating; (3) Al coating (Van Steenkiste et al. 2002).

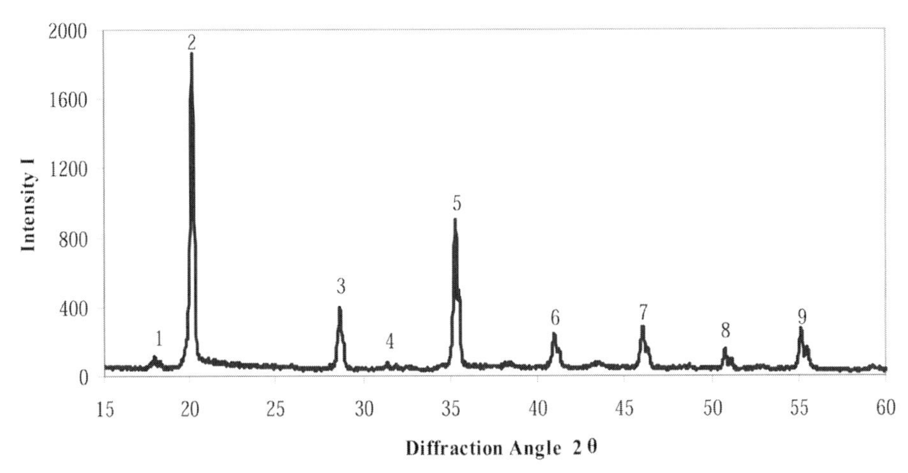

Fig. 5.9 XRD spectra of Al–W–Zn coating. 1,3,8,9 – $Al_{0.71}Zn_{0.29}$; 2, 4, 5, 6, 7 – WAl_{12}.

10–20 units, proving the assumption that mechanical alloying reactions occur (Fig. 5.9).

5.4 Basic Mechanisms

Based on the experimental data presented in the previous chart, it is important to discuss three primary issues: (i) the basis for the application of Arrhenius' law in the phenomenological analysis of GDS processes, (ii) the behavior of powder mixtures during GDS, and (iii) the characteristics of shear localization and mechanical alloying in GDS coating structure formation.

An expression for the deposition efficiency and average single pass coating thickness may be obtained using the Arrhenius equations (5.1) and (5.2). This appears to be reasonable because, according to Eq. (5.3) (Dykhuizen and Smith 1998), the particle velocity within the powder jet is a function of the gas temperature:

$$V_p = (M-1)\sqrt{\frac{\gamma_c R T_0}{1+[(\gamma_c-1)/2]M^2}} \tag{5.3}$$

Here T_0 is the initial gas temperature (in the convergent part of the nozzle), M is the Mach number, and γ_c is the ratio of specific heats or isentropic coefficient (Stoltenhoff et al. 2002). For monatomic gases, γ is 1.66, and for diatomic gases γ is typically 1.4. (Air is typically modeled as a diatomic gas because it is primarily composed of nitrogen and oxygen.)

The modeling results and analytic equations based on the isentropic gas flow model that were developed by Dykhuizen and Smith (1998) allow for a rough estimation of the particle velocity. This analytical equation (5.3) demonstrates the dependence of particle velocity on gas temperature. As Dykhuizen et al. describe, this equation may be used to approximate the particle velocity only when an optimal nozzle shape is used. Further, this relationship does not take into account particle parameters. It is in fact quite difficult to obtain analytical solutions using equations whose main variables are the drag coefficient, particle density, and particle size. Equation (5.3), however, provides a means for the preliminary analysis of the dependence of particle velocity on the temperature and type of the propellant gas (RT_0).

The collision contact time t_c can be approximated by a simplified form of Eq. (2.26):

$$t_c \approx \frac{d}{V_p} \tag{5.4}$$

where d is the diameter of the incoming particle. The severe deformation of particles due to impact is described by the shear strain γ which may be approximated using microstructure analysis. The shear strain rate may be calculated as

$$\dot{\gamma} \approx \frac{\gamma}{t_c} \tag{5.5}$$

One can find RT_0 from Eq. (5.3) as

$$RT_0 = f(\gamma_c, M) V_p^2 \qquad \text{where} \quad f(\gamma_c, M) = \frac{1 + \frac{\gamma_c - 1}{2} M}{(M-1)^2 \, \gamma_c}$$

Finally, Eq. (5.2) may be modified by taking into account (5.4) and (5.5):

$$t = t_0 \exp\left(-\frac{Q_t}{f(\gamma_c, M) \, d^2 \left(\frac{\dot{\gamma}}{\gamma}\right)^2} \right) \tag{5.6}$$

This relationship suggests that the overall effectiveness of the GDS process is determined by the relation (or contact time), which as presented in the previous chapter, characterizes the process of adiabatic shear bands formation at the particle interface. Wright (1987) and Wright and Batra (1985) show that under extreme loading conditions various materials exhibit narrow bands of intense plastic deformation, termed shear bands. The thermomechanical mechanism of shear bands formation is associated with shear localization during high strain rate loading in rate-sensitive materials such as polycrystalline metals or granular materials (Hu and Molinari 2004). Under this mechanism, thermal softening is thought to overcome hardening effects in the material, and to push forward the shear band formation process. As shown by Wright (1987), Wright and Batra (1985), Walter (1992), and Zerilli and Armstrong (1997), a flow law having an Arrhenius dependence on temperature may be used to model shear localization and shear band phenomena in thermoviscoplastic materials. Flemming et al. (2000) completed the first comprehensive analytical study of shear localization with the Arrhenius flow law, yielding a one-dimensional formulation based on plastic strain-rate function ("flow law") analysis. This strain-rate function, $\dot{\gamma}$, depends on both temperature and stress. Physically speaking, the nature of the plastic flow governs shear bands formation, and its mathematical representation is the source of nonlinearity in the problem (Flemming et al. 2000). Theoretical models based on the microstructure and the physics of deformation employ the thermal activation of plastic flow, leading to an Arrhenius dependence on temperature in accordance with Kochs et al. (1975), Dodd and Bai (1989), and Zerilli and Armstrong (1997). Such models are used in numerical studies by Wright (1987); however, in many analytical studies the flow laws are approximated so as to simplify calculations. In particular, the following exponential flow law by Flemming et al. (2000) is often used:

$$\dot{\gamma} = \dot{\gamma}_0 \tau^N e^{-\frac{Q}{RT_D}} \tag{5.7}$$

In this equation T_D is the deformation temperature, Q is the activation energy, N is the power coefficient, and τ is the shear stress in the area of shear localization. The main feature of (5.7) is the Arrhenius dependence of shear strain on temperature. The form of (5.7) has been chosen by Flemming et al. (2000) to allow two types of stress dependence: $N = 0$ and $N > 0$. In comparing the forms of (5.6) and (5.7) it can be seen that macroscopic characteristics of the GDS process such as the average single pass thickness, t, or the deposition efficiency, χ_D, depend on the kinetics of adiabatic shear bands formation which is described by means of the Arrhenius flow law. Indeed, in order to better understand the role of shear localization as the main coating formation process during particle impact loading in GDS, the relationship between Eqs. (5.6) and (5.7) needs to be further examined.

The detailed analysis of the flow characteristics of gas-particle suspensions in a powder-laden jet seems to be important for the evaluation of particle–particle collisions and particle–turbulence interactions. These processes were studied both for diluted (Senior and Grace 1998) and concentrated gas–solid suspensions (Eskin and Voropaev 2004). Despite the relatively high gas velocities (300–900 m/s) in these "fast fluid bed units (Senior and Grace 1998), the concentration of suspended solids is neither uniform along the nozzle axis, nor over the cross-section of the flow. High particle concentration generally forms near the nozzle walls where particles move with lower velocities. Clusters of particles may also intermittently form and disintegrate in the diluted suspension flows. Thus, the probability of particle cluster formations may be evaluated through the differences in axial particle velocities ΔV_i. The gradient $\Delta V_{(1-50)} = V_1 - V_{50} =\sim 600$ m/s, where V_1 and V_{50} are the velocities of particles having sizes of 1 μm and 50 μm, respectively. In fact, even a small difference in particle size results in a velocity gradient of about 100–200 m/s. The scheme of particle clustering for the case of spraying bimodal powder mixtures is shown in Fig. 5.2(b).

As a first step in attempting to model the GDS process two regions of the powder-laden jet must be taken into consideration:

i. the free jet region (whose length is nearly equal to the stand-off distance); and
ii. the region of the normal shock wave layer.

The average concentration of particles rebounding from the substrate is high in the normal shock wave layer. This leads to an increase of particle concentration in clusters moving through the normal shock wave front. Two conclusions may be drawn:

i. the velocity of the cluster depends on the particles rebound process; and
ii. the coating formation in the case of particle clusters impingement with a substrate is similar to powder shock compaction.

In this case, particles having low velocities and high weights play the role of shot balls in the powder shock consolidation process. These particles are distributed within the matrix of small particles (Fig. 5.7). Therefore, the impact powder microforging process is realized. The ratio of the particle sizes, weights, and concentration governs the efficiency of the powder microforging impact, resulting in bonding and the densification of powder layers.

An XRD spectrum of an Al–W–Zn coating is shown in Fig. 5.9. After GDS, the observed XRD peaks show new phases which are a result of deformation enhanced alloying. The mechanical alloying processes (see the review by Zhang (2004)) are known to develop due to intensive particle deformation during high energy mechanical milling. Thus, it can be assumed that the main reason for the presence of mechanical alloying processes in GDS is the severe deformation of particles and shear localization at the particles' interface.

5.5 Concluding Remarks

Experimental, phenomenological, and microstructural analyses of air LPGDS of the powder mixtures injected in the diverging part of a supersonic nozzle have been performed. The results are summarized as follows:

- the GDS of powder mixtures results in the intensive process of powder microforging and the severe deformation of particles;
- the main characteristics of GDS – such as deposition efficiency and average single pass thickness – depend on the kinetics of adiabatic shear band formation following the Arrhenius flow law;
- application of powder mixtures results in mechanical alloying during particle impact and consolidation;
- the particle cluster formation mechanism in the powder-laden jet governs the impact powder consolidation features of LPGDS.

In short, then, it is necessary to properly diagnosis the parameters of GDS. Further, sufficient analysis of the powder-laden jet structure will help to define the influence of pressure on its characteristics and properties. This will be undertaken in the following chapter.

6
Diagnostics of Spray Parameters: Characterization of the Powder-Laden Jet

6.1
General Relationships

Widespread adoption of a cold gas dynamic spray technology requires the application of a variety of coating materials, each one specifically suited to particular applications. For this reason it is necessary to determine the dependence of the microstructure of spray coatings on the operating conditions of the gas dynamic spray system which is of great practical interest. To obtain high quality coatings, the spray technology parameters must be selected carefully, and due to the large variety of process parameters, much trial and error goes into optimizing the process for each specific coating and substrate combination. Consequently, a complete model of the GDS coating process should be developed. This model should include three distinct sub-processes: spray parameters such as particle size, temperature, velocity, and impact points; particle impact, deformation and bonding; and the coating microstructure, including shock compaction and adiabatic shear bands formation, etc.

In the cold GDS process, particles, in metallic or nonmetallic form, are fed into a carrier gas where they are heated, accelerated, and propelled at high velocity onto the surface to be coated (Fig. 6.1). Particles land on the solid surface where they deform, and agglomerate to create a thin layer. It is well known that the introduction of new coating materials is both time consuming and costly. To date, GDS spray parameters are being optimized merely on the basis of trial and error, from which empirical relationships are derived. Since a very large number of parameters are involved (e.g., gas type and temperature; particle size and velocity; substrate material and temperature) this approach is laborious and expensive, and must be repeated for each material being sprayed. The development of theoretical or computational models which can predict the microstructure of coatings could greatly reduce the cost of this development. Ideally, these models will allow operators to customize coating properties to meet the requirements of particular applications without having to do extensive experimentation. Indeed, these models will also lead to improved designs of GDS systems which optimize the technological parameters of GDS processes. Indeed, the modeling of low-pressure GDS process will

Introduction to Low Pressure Gas Dynamic Spray Roman Gr. Maev and Volf Leshchynsky

ISBN: 978-3-527-40659-3

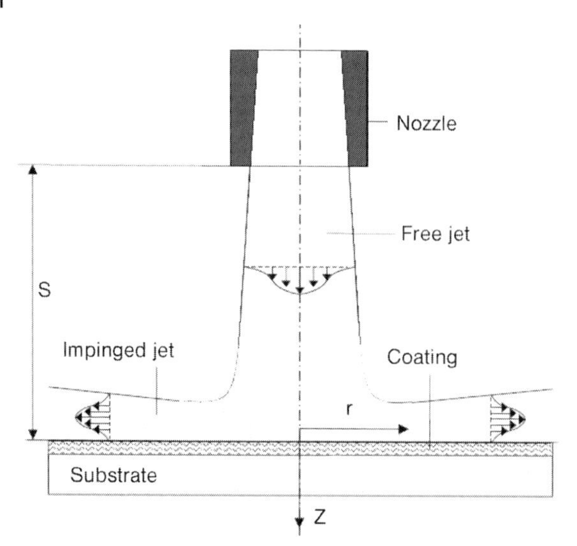

Fig. 6.1 Particle-laden jet scheme.

promote the widespread adoption of a cold gas dynamic spray technology in expanded industrial environments.

As suggested earlier, a complete description of GDS includes modeling three distinct process: particle acceleration (the main variables being particle size, density; gas type, pressure and temperature), particle impact deformation and bonding (the main variables being particle velocity, density, temperature, particles distribution in the gas-powder jet, stress–strain state at an interface); and the powder coating microstructure formation including adiabatic shear banding, particle consolidation and mechanical alloying (the main variables being shock compaction parameters of different powder mixtures, characteristics of a heat generation due to impact, diffusion constants, etc.). Specifically this model consists of several complementary parts and should include sub-models that simulate: (a) gas flow and particle acceleration; (b) impact, deformation, and bonding of particles on the substrate; and (c) shock compaction (densification) and consolidation of powder layers. The last two parts of GDS process characterization have been analyzed in Chapters 3 and 4.

There are many industrial processes similar to gas dynamic spraying, which involve jets impinging on a solid surface. Such examples include explosive welding, jet mixing systems, water jet cutting, and shrouding systems. Many of these are discussed extensively in books by Schlichting (1979), Abramovich (1963), and Hinze (1975). For the specific problem of turbulent free jets and compound jets consisting of more than one flow stream, the flow structure is well understood, with close agreement being demonstrated between theory and practice by Rajaratnam (1976). However, present literature does not include much work on compound jets containing flowing powder.

6.1.1 The Governing Equations of Single-Phase Turbulent Flow

One feature of turbulent jet flow is that momentum, heat, and mass are transferred at rates much greater than those of laminar flow (where molecular transport processes take place via viscosity and diffusion) (Sabersky et al. 1971). The conservation equations used for turbulent flows are obtained from those of laminar flows using a time averaging procedure commonly known as Reynolds averaging (Rajaratnam, 1976). When the flow is turbulent, the characteristic variables of pressure, velocity, and temperature may vary with both space and time.

The mathematical model used in simulation is based on the Navier–Stokes system of differential equations with the Reynolds method of averaging the time-dependent equations, together with the standard k–ϵ turbulence model. The time-averaged, governing equations for turbulent flow are expressed as follows (Rajaratnam, 1976):

- Conservation of mass:

$$\frac{\partial}{\partial x_j}(\rho_g v_j) = 0 \tag{6.1}$$

where ρ_g is the density of the gas, and v_j is the velocity vector in the jth direction.

- Conservation of momentum:

$$\frac{\partial}{\partial x_j}(\rho_g v_i v_j) = -\frac{\partial p}{\partial x_j} + \frac{\partial \tau_{ij}}{\partial x_j} + \rho_g g_i \tag{6.2}$$

where τ_{ij} is the stress and g_i is the gravitational acceleration.

The stress is given by

$$\tau_{ij} = \left[(\mu + \mu_t)\left(\frac{\partial v_i}{\partial x_j} + \frac{\partial v_j}{\partial x_i}\right)\right] - \frac{2}{3}\mu_t \frac{\partial v_i}{\partial x_i}\delta_{ij} \tag{6.3}$$

where μ is the molecular viscosity and $\delta_{ij} = 1$ for $i = j$; otherwise $\delta_{ij} = 0$. Also, μ_t is the turbulent viscosity given by

$$\mu_t = \rho_g C_\mu \frac{k_t^2}{\epsilon} \tag{6.4}$$

Where $C = 0.09$ is a constant, k_t is the kinetic energy of turbulence, and ϵ_t is the dissipation of kinetic energy of turbulence, which will be defined in the k–ϵ turbulence model to follow.

6.1.2
The k–ϵ Model for Turbulent Flows

The modeling of turbulent flows requires appropriate methods to describe the effects of the unstable fluctuation of the velocity and scalar quantities within the basic conservation equations. Since numerical simulations of the turbulence have been thoroughly studied, the scope of this description is not concerned with model development, but focuses only on adopting a reliable turbulence model for jet flow problems.

The standard k–ϵ model of turbulence is generally used to close the system of conservation equations described in Eqs. (6.1)–(6.4). Two additional conservation conditions include the kinetic energy of turbulence k_t and its dissipation ϵ_t. In accordance with FLUENT[1], these are

$$\frac{\partial}{\partial x_i}(\rho v_i k) = \frac{\partial}{\partial x_i}\left(\frac{\mu_t}{\sigma_k}\frac{\partial k}{\partial x_i}\right) + G_k + G_b - \rho_g \epsilon_t \tag{6.5}$$

and

$$\frac{\partial}{\partial x_i}(\rho v_i \epsilon) = \frac{\partial}{\partial x_i}\left(\frac{\mu_t}{\sigma_k}\frac{\partial \epsilon_t}{\partial x_i}\right) + C_{1\epsilon}\frac{\epsilon_t}{k_t}(G_k + G_b) - C_{2\epsilon}\rho_g\frac{\epsilon_t^2}{k_t} \tag{6.6}$$

Here,

$$G_k = \mu_t\left(\frac{\partial v_j}{\partial x_i} + \frac{\partial v_i}{\partial x_j}\right)\frac{\partial v_i}{\partial x_j} \tag{6.7}$$

$$G_b = -g_i \frac{\eta_t}{\rho_g \sigma_k}\frac{\partial \rho_g}{\partial x_i} \tag{6.8}$$

with $C_{1\epsilon}$, $C_{2\epsilon}$, σ_k, and σ_ϵ being the empirical constants, σ_h the turbulent Prandtl number, ${}_tC_p/k_t$, G_k the rate of production of kinetic energy of turbulence, and G_b the generation of turbulence due to buoyancy (FLUENT users guide).

6.1.3
Particle Dynamics in Gas Flow

The effective characterization of the GDS particle-laden jet provides a means for determining the drag force per unit particle mass, drag coefficients, and the effect of friction on the particle velocity for various particle-to-gas mass flow ratios. Naturally, these parameters must defined with respect to particle flow, as the behavior of particles suspended in turbulent flow depends on the properties of both the particles and the flow. Turbulent dispersion of both the

1 FLUENT (1996) is a software package which models turbulent gas-particle flow. It utilizes a finite difference numerical algorithm based on a specified volume control approach.

particles and the carrier fluid can be handled via energy analysis for definite particle size distributions. The momentum transfer during the interaction between the gas and particle phases has been thoroughly investigated and many criteria have been isolated. However, due to the inherent computational complexity that exists in particle and gas flow analysis, Streeter (1961) suggests that the uncertainty of current prediction models is still large. In fact, in order to define the properties of a gas-particle mixture, the volume must be large enough to contain sufficient particles to achieve a stationary average. For this reason the particles cannot be treated as a continuum in a flow system of comparable dimensions.

A two-phase flow problem involving a dispersed second phase may be solved through the addition of a second transport equation. The trajectory of a dispersed phase particle is solved by integrating the force balance on the particle in a Lagrangian reference frame (following the particle coordinate). Usually, investigators treat a solid phase either as a discrete system (Triesch and Bohnet 2001) or as a continuum (Gidaspow 1994). For computing dilute flows researchers (e.g., Triesch and Bohnet 2001) and Oesterle and Petitjean 1993) often use a discrete method where separate particle trajectories are calculated through a Lagrangian approach while particle–particle collisions are treated as a random process using the Monte Carlo algorithm. It is assumed that a particle moves in a cloud formed by other particles. The motion through this cloud is accompanied by collisions with particles forming the cloud. The continuum approach of Gidaspow (1994) and Louge et al. (1991) is based on kinetic theory. According to this theory, particles are considered to be chaotically moving gas molecules, and this granular gas is characterized by its granular pressure, temperature, conductivity, and the flow viscosity. Here the flow viscosity (which is calculated using the theory of inhomogeneous gases) accounts for the stresses caused by the chaotic motion of the particles.

The force balance equation equates the particle inertia with the force acting on the particle, and can be written as

$$\frac{dv_{p,i}}{dr} = f_D(v_{\omega,i} - v_{p,j}) + \frac{g_i(\rho_p - \rho_\omega)}{\rho_p} + F_i \tag{6.9}$$

where $f_D(v_{\omega,i} - v_{p,j})$ is the drag force per unit particle mass and

$$f_D = \frac{18\mu}{\rho_p d_p^2} \frac{C_D Re}{24} \tag{6.10}$$

where Re stands for the relative Reynolds number, which is defined as

$$Re = \frac{\rho_\omega d_p |v_{p,i} - v_{\omega,i}|}{\mu} \tag{6.11}$$

The drag coefficient, C_D, is a function of the relative Reynolds number and is of the following general form (Walsh 1975):

$$C_D = \alpha_1 + \frac{\alpha_2}{Re} + \frac{\alpha_3}{Re} \tag{6.12}$$

where α_1, α_2, and α_3 are constants which apply over several ranges of Re. Integration of Eq. (6.9) yields the velocity of the particle at each point along the trajectory, with the trajectory itself predicted by

$$\frac{dx_i}{dr} = v_{p,i} \tag{6.13}$$

To adequately predict the trajectories of the dispersed phase the basic force balance equation (6.9) must be solved in each coordinate direction. However, this equation incorporates additional forces F_i in the particle balance which can be important under special circumstances such as interparticle friction and friction between the particles and the walls. Thus, in order to solve the system it is necessary to incorporate the friction force equations.

Since the friction of the particles is determined by the shear stress at the wall, gas-particle flow may be treated as continuous flow and may be described in terms of fluid dynamics.

Experimental studies of a gas flow containing monodisperse particles in a round tube by Tsuji et al. (1984) revealed a considerable effect of friction on the particle velocity for various particle-to-gas mass flow ratios. It is therefore of great importance to account for the effects of friction at the gas-particle jet both inside and outside the nozzle.

To describe the supersonic gas-powder flow behavior in a De Laval nozzle in GDS processes, the fundamental results of computational fluid dynamics (CFD) were used. Further, analyses of gas and particle dynamics were undertaken using the principles of kinetic theory that were put forth by Gidaspow (1994) and Louge et al. (1991) and further developed by Eskin et al. (2003).

6.2 Gas Flow and Particle Acceleration

A single stage of the gas flow and particle acceleration may be divided into three periods (Fig. 4.1):

1. gas and powder flow and mixing;
2. movement and acceleration of particles in the divergent part of the nozzle;
3. movement of particles in free jet area (including rebound from the substrate).

Each of these processes has its own features and may be characterized by several fundamental parameters. The analysis of these characteristics through physical and numerical modeling allows for the optimization of these values. As shown in Fig. 6.1, the gas and particle velocities change at the each period of acceleration. One can also see that there is a significant difference between gas and particle velocities due to the variation of the drag force.

As shown in Chapter 4, the volume solid content of the gas-particle jet in classical GDS is in the range of 10^{-5}–10^{-6}, while the volume content in low-pressure GDS system is in the range of 10^{-4}–10^{-5}. For this reason parameters of gas-powder flow vary considerably and are strongly dependent on the particular GDS system. For example, in low-pressure systems interparticle friction plays an increasingly significant role. However, the lack of modeling and experimental data does not permit the exact evaluation of such specific parameter dependencies.

Researchers have historically calculated nozzle accelerations without taking into account particulate friction (Rudinger 1980). This can in fact be justified if the solid content is low and/or the ratio of the nozzle's length to its diameter is relatively small. Acceleration calculations for the case of high volume solid content were made by Eskin and Voropaev (2004). Here the authors describe, in detail, how two frictional models may be applied to acceleration nozzles. In the first method the friction factors for solids are employed. Using a method often employed in engineering calculations of pneumatic conveying, a fluid-like behavior of the particulate phase is assumed. Unfortunately, the application of this method leads to discrepancies between calculated and measured pressure gradients that have been shown to reach and even exceed 40%. In the second model the frictional forces between solids is taken into account by assuming that the particles acquire radial velocity components due to noncentral collisions with particles of various sizes (Eskin et al. 1999, 2003, 2001). This radial motion of particles leads to particle-wall collisions which give rise to partial losses in the particle axial momentum components. This momentum loss is considered to account for particulate friction. The major disadvantage of this model is that it does not take into account collisions between particles of equal size. However, when this method is being applied to GDS processes using powders consisting of varying particle size and content this disadvantage is assumed to be minimal.

Eskin and Voropaev (2001) provide detailed analysis of flow acceleration within a jet mill nozzle. Perhaps the most interesting finding of this investigation is that the ratio between the energy transferred from the gas to the particles and that to which acts to accelerate the mixture is dependent on both particle size and solids loading. In fact, the authors demonstrate through thermodynamic analysis that this ratio very rapidly increases with a decrease in particle size, and significantly increases with an increase in solids loading. The most important results of particle acceleration processes modeling in De Laval nozzles will be discussed below from this point of view.

6.2.1 Computational Fluid Dynamics (CFD)

Numerical simulations are often very helpful in analyzing the GDS process. As such, a short review of numerical simulation results is presented by Jen et al. (2005). Numerous GDS characteristics such as nozzle geometry, processing parameters, and spray particle parameters have been examined using CFD code. Through this analysis, it has been found that the primary processing parameters that influence the particle velocity are the carrier gas type, operating pressure, and temperature. As for nozzle geometry, the expansion ratio and divergent section length of the spray gun nozzle also have a significant effect on particle velocity. Moreover, the density, size, and morphology of powder play a notable role. The effects of these main parameters are sufficiently summarized by an equation that may be obtained through the nonlinear regression of the simulated results for particle velocity (Jen et al. 2005). Li et al. (2005) have conducted numerical simulations for two-phase gas–solid flows. Here the airflow field is obtained by solving three-dimensional Navier–Stokes equations with a standard k–ϵ turbulence model and nonequilibrium wall function. The second phase, the coating powder, is thought to consist of spherical particles that are dispersed within the continuous phase (the air). In addition to solving the transport equations for the air, the authors calculate the trajectories of the particles by solving the particle motion equations using a Lagrangian method.

As shown previously, high-pressure GDS process is characterized by low particle-gas flow ratios. Several recent CFD analyses of gas and particle dynamics in GDS processes have been performed by McCune et al. (1995), Dykhuizen et al. (1998, 1999, 2003), and Sakaki and Shimizi (2001). The gas flow model was first introduced by Dykhuizen and Smith (1989), Dykhuizen and Smith (1998), Dykhuizen et al. (1999), and Dykhuizen and Neiser (2003) who modeled the internal and external flow of a supersonic nozzle in the GDS process. Over the years, in order to optimize system and nozzle design, many studies have focused on the behavior of flying particles. In this case it is assumed that the gas flow conditions can be calculated without considering feedback from the powder flux.

The effect of Mach number (speed of sound) on particle velocity has been investigated with respect to an isentropic gas flow model. In this work, Dykhuizen and Smith (2003) assume that the gas flow is isentropic (adiabatic and frictionless) and one-dimensional. Thus, heat and friction losses are not considered. Under these simplified conditions, the changes of state are functions of the local Mach number and the isentropic coefficient or specific heat-capacity ratio γ (Stoltenhoff et al. 2002):

$$\frac{p_g}{p_0} = f(M,\gamma); \ \frac{\rho}{\rho_0} = f(M,\gamma); \ \frac{T}{T_0} = f(M,\gamma) \tag{6.14}$$

The subscript 0 represents the upstream stagnation condition of the gas.

In the isentropic gas flow model, sonic conditions for flow gas may only be obtained for sufficient stagnation gas pressures (Dykhuizen et al. 1999). After the sonic condition is obtained, increasing the pressure does not result in an effective increase in gas velocity upstream of the throat. In this model, the gas velocity v_g can be expressed as

$$v_g = Mv_s = M\sqrt{\gamma RT}M_w \tag{6.15}$$

where M is the local Mach number, v_s is the speed of sound, and γ is the ratio of specific heats. For monatomic gases γ is 1.66, and for diatomic gases γ is typically 1.4. Also, R is the gas constant (J/kmol K), T is gas temperature, and M_w is the molecular weight of the gas. The local Mach number M only depends on the inner form of the nozzle, the calculation of which is presented in detail by Dykhuizen and Smith (1998), Dykhuizen et al. (1999), and Dykhuizen and Neiser (2003). Typical cold-spray geometry includes a converging–diverging nozzle, which is shown schematically in Fig. 4.1.

In one-dimensional numerical (Wallis 1969) models concerning particle acceleration the particle velocity is calculated by defining the drag force on a single particle in fluid flow on the basis of the following equation:

$$m\frac{dv_p}{dt} = mv_p\frac{dv_p}{dx} = \frac{C_D A_p \rho_g (v_g - v_p)^2}{2} \tag{6.16}$$

It can be seen from (2.23) that the gas pressure does not affect the gas velocity directly. However, the drag coefficient C_D is a function of Re, and the particle Mach number M_p, i.e., $C_D = f(M_p, Re_p)$ (Henderson 1976). Re is in turn a function of the gas density ρ_g which itself depends on gas pressure

$$p_g = \rho_g RT$$

Alkhimov et al. (2001) obtained an empirical equation which includes the particle velocity, the gas velocity, and the gas pressure for nitrogen (which is often used as a processing gas):

$$v_p = \frac{v_g}{1 + 0.85\sqrt{\frac{d_p}{x}}\sqrt{\frac{\rho_s v_g^2}{p_0}}} \tag{6.17}$$

Here v_p is the particle velocity, p_0 is the nitrogen supply pressure measured at the entrance of the nozzle, ρ_s is the particle density, d_p is the particle diameter, and x is the axial position. If the gas velocity is replaced with Eq. (6.15), the correlation between the particle velocity, the gas temperature and pressure can be rewritten as:

$$v_p = \frac{1}{\frac{1}{M}} \sqrt{\frac{M_{N_2}}{\gamma RT} + 0.85\sqrt{\frac{d_p}{x}}\sqrt{\frac{\rho_s}{\rho_0}}} \tag{6.18}$$

where M_{N_2} is the molecular weight of the nitrogen gas.

The magnitude of particle acceleration and particle velocity can be obtained from the isentropic gas flow model and the one-dimensional numerical particle acceleration model. Unfortunately, some complex phenomena such as shock waves, flow separation, and the velocity distribution of in-flight particles cannot be adequately analyzed. Even more, there is little experimental data available about the velocity distribution of in-flight particles. In the paper of Wu et al. (2005), a professional velocity measuring device is used, and experimental velocity measurement results are presented. The strong particle velocity dependence on pressure and velocity distribution within the nozzle suggests that it is necessary to account for gas pressure and nozzle geometry along with gas type.

Nozzle geometry is of utmost importance in both cold gas dynamic spraying and high-velocity oxygen-fuel (HVOF) processes. For this reason GDS research has in recent years been focused on nozzle modeling and design. For example, Sakaki and Shimizu (2001) carried out both numerical simulation and experiments to investigate the effect of the entrance geometry of the gun nozzle on the HVOF process. (The results of this study may be easily applied to the cold spray method.) A similar research project concerning wedge-shaped supersonic nozzles was also conducted by Alkhimov et al. (2001). Here the authors show that for a particular nozzle and particle type it is possible to adapt the GDS parameters in such a way as to maximize the particle velocity that may be achieved.

Dykhuizen and Smith (1998), Dykhuizen et al. (1999), and Dykhuizen and Neiser (2003) employed the CFD program FLUENT to simulate the use of nitrogen and helium to accelerate copper particles having diameters varying from 5 μm to 25 μm and inlet conditions of $P_0 = 2.5$ MPa and $T_0 = 673$ K. They found that particle velocity is a controlling factor that can determine the properties of bonding with the substrate. The carrier gas helium was found to be 2.5 times faster than nitrogen in the supersonic nozzle. Furthermore, smaller particles were shown to travel faster than larger particles, and an increase in either gas pressure or temperature contributed to an increased particle velocity.

The primary goal of this book is to adequately describe the main physical principles of a portable low-pressure CGDS system. Further, it is to characterize which attributes make this technology suitable for use within an industrial environment. Indeed this knowledge holds the key to such cross-disciplinary applications of GDS as cold surface coating technology. Before addressing the potential for new low-pressure CGDS coating devices, however, it is necessary to understand the particle transport process within the supersonic gas stream from the inlet of the nozzle to the substrate. Naturally, the key aspect of successful low-pressure CGDS powder coating processes is the bonding on the substrate and consolidation of the powder layer. In order to bond powder particles to the substrate upon impact, the particles need to attain enough momentum. In short, the velocity in the powder-laden jet must be greater than a critical velocity for a particular material and particle size. Below this critical velocity particles rebound from and ultimately erode the substrate.

Although helium, due to its light weight, is the best carrier gas for particle acceleration (Voyer et al. 2003), it is very expensive. For this reason alternative compressed gases such as nitrogen or air are commonly used. The critical gas velocity (i.e., the velocity of particle consolidation and dense coating formation) is material dependant and is affected by the density and elastic–plastic properties of the sprayed powder material, its composition and particle sizes (Stoltenhoff et al. 2002, Vlcek et al. 2003). As shown by Van Steenkiste et al. (1999) and Van Steenkiste and Smith (2003), the critical velocity for GDS of Al-based powders accelerated by the compressed air, is in the range of 400–500 m/sec. Moreover, the use of relatively large particles in air GDS processes allows for the minimum kinetic energy requirement for bonding to be readily attained.

Maev et al. (2006) have demonstrated that a portable GDS machine with a relatively low compressed air pressure of about 5–8 bars can be used to achieve critical particle velocities for Al and Ni powders. Further, Maev et al. (2004) show that greater nozzle durability and simpler powder hopper designs may be achieved if a radial injection method of powder insertion is used. The acceleration of particles in De Laval nozzles at the air pressures in the range of 0.2–1.0 MPa is analyzed by Eskin et al. (1999), Eskin and Voropaev (2001), Eskin and Kalman (2002), Eskin and Voropaev (2004). These results may be used for analysis of the LPGDS process.

6.2.2 An Engineering Model with Particle Friction

In accordance with Eskin et al. (2003), the force acting on ith fraction of particles may be treated as the sum of three forces:

$$f_i = f_{Di}^{(g)} + f_i^{(c)} + f_i^{(\mathrm{fr})} \tag{6.19}$$

Here, $f_{Di}^{(g)}$ is the drag force, which is defined by Eq. (6.10), and $f_i^{(c)}$ is the force arising from the collisions of particles having different sizes (particle rotation is neglected). Thus,

$$f_i^{(c)} = m_i \sum_{j=1}^{n} \frac{\langle \Delta v_{ij} \rangle}{t_{ij}} \tag{6.20}$$

where

$$\langle \Delta v_{ij} \rangle = \frac{m_j}{m_i + m_j} (v_j - v_i)|v_j - v_i| \frac{1 + k_n}{2} \tag{6.21}$$

is the change in the axial velocity v_i as a result of a collision of a particle within the ith fraction with one from of the jth fraction (averaged over all the possible directions of eccentric collisions). k_n accounts for the change in particle velocity due to the inelasticity of collisions, and

$$t_{ij} = \frac{4m_j}{\rho_g (d_i + d_j)^2 \rho_p \epsilon_j |v_j - v_i|} \tag{6.22}$$

is the mean free time between interparticle collisions.

Having developed the mathematical basis for particle–particle friction it is now beneficial to derive an equation with which to calculate the forces arising from particle-on-wall friction $f_i^{(\mathrm{fr})}$. These forces are believed to be related to the radial motion of particles that results from the eccentric collisions between particles in the ith and jth fractions. The radial velocity v_{rij} of a particle within the ith fraction after collision with a particle of the jth fraction is

$$v_{rij} = \frac{m_i}{m_i + m_j} \frac{(1 + k_n)}{4} |v_j - v_i| \tag{6.23}$$

where, according to Jenkins (1992), k_n may be used to represent the loss of kinetic energy due to the inelasticity of collisions, $\delta E = 1 - k_n$. For most powders $k_n = 0.4 - 0.5$; that is, 75–84% of the energy is lost because of inelastic collisions.

The equation for the radial projection of the drag force is

$$m_i = \frac{v_{rij} - v_{rij}^{(f)}}{t_ij} = F_{ij}^{(r)} \tag{6.24}$$

where the radial projection of the drag force can be calculated by the equation

$$F_{ij}^{(r)} = \frac{1}{8} p_g C_{Di} \pi d_i^2 |v_g - v_i| v_{rij}^{(m)} \tag{6.25}$$

The mean radial velocity of a particle moving in a polydisperse medium can be calculated using the averaging relationship

$$v_{ri} = t_i \sum_{f=1}^{n} \frac{v_{rij}}{t_{ij}}, \; t_i = \frac{1}{\sum_{j=1}^{n} \frac{1}{t_{ij}}} \tag{6.26}$$

Here, $v_{rij}^{(m)}$ is the mean radial velocity of a particle of the ith fraction in the time interval t_{ij}. (Note that t_{ij} represents the time from the instant it collides with a particle of the jth fraction until it collides against the nozzle wall). On the basis of the assumption that the radial velocity decreases linearly with time, the mean radial velocity of the particles in the time interval t_{ij} can be defined as $v_{rij}^{(m)} = 0.5(v_{rij} + v_{rij}^{(f)})$, where $v_{rij}^{(f)}$ is the radial velocity of a particle of the ith fraction that has collided with a particle of the jth fraction at the instant it is colliding against the wall. Finally, t_i is the mean free time of the particle in the polydisperse medium.

Based on these parameters, Eskin et al. (2003) obtained the following equation for the average radial velocity of the particles:

$$v_{ri}^{(m)} = \frac{v_{ri} + v_{ri}^{(f)}}{2} = v_{ri}\frac{1}{1+\chi_i} \tag{6.27}$$

where,

$$\chi_i = \frac{3}{8}C_{Di}\frac{\rho_g|v_g - v_i|t_i}{\rho_s d_i} \tag{6.28}$$

is the mean drag coefficient for the radial motion of the particles in the time internal t_i.

The force arising from particle-on-wall friction $f_i^{(\mathrm{fr})}$ may be determined by using the "colliding layer" estimation. Any particle of this layer is assumed to collide against the wall if its velocity vector (after a collision with another particle) is directed toward the wall. The thickness of this layer for particles of the ith fraction is

$$h_i = v_{ri}^{(m)} t_i \tag{6.29}$$

The friction force equation is given by

$$f_i^{(\mathrm{fr})} = -\frac{1}{2}\psi\frac{m_i v_i}{t_i}\left[1 - \left(\frac{2h_i}{D} - 1\right)^2\right] \tag{6.30}$$

where the coefficient ψ $(0 \leq \psi \leq 1)$ takes into account that the velocity loss due to the wall collision:

$$\delta u_i = -\psi u_i \tag{6.31}$$

The model developed by Eskin et al. (2003) enables one to both calculate the loss of energy due to friction for nozzles of various designs and estimate the scaling factor. Although this model is developed for rarefied flows, it is consistent with kinetic theory. However, in this case mean radial particle velocities replace particulate medium temperatures as the cause of the chaotic particle motion. Indeed, it is clear from the analysis presented above that chaotic motion is caused by interparticle collisions, which are induced by the polydispersity of the flow. For this reason the application of polydispersive mixtures in GDS processes may contribute to an overall improvement in technology.

6.3 Calculated Data and Discussion

6.3.1 Simulation of Gas-Particle Flow in the Nozzle

Consider for a moment the gas-particle flow parameter data that has been presented by various authors based on the models described above. A comparison of the deposition characteristics of classical and low-pressure GDS techniques is shown in Fig. 4.3. Here a considerable difference in the solid volume concentration of the gas-powder jet (10^{-4}–10^{-6}) can be observed for each GDS system. The specific influence that particle content has on particle velocity in a powder-laden jet, however, has not yet been estimated for real GDS conditions. On the other hand, some simulation results for heavily laden gas-particle jets have been obtained which take into account both particle collisions and wall friction effects. They are presented in a series of papers by Eskin et al. (1999, 2001, and 2002). In this case the solid volume fraction of the gas-particle jet is in the range 10.3–10.1. The simulation results from Eskin et al. (1999), along with the experimental data of Mebtoul et al. (1996), have been used to construct a plot of exit particle velocity as a function of solid volume concentration in gas-particle jet (Fig. 6.2(a)). One can see that the dependency $v_p = f(\epsilon)$ fits the power approximation (Fig. 6.2(b)) with a veracity of 0.98–0.99.

Some velocity measurements for particles of size 120 μm were made by Mebtoul et al. (1996) in an air jet having a stagnation pressure of 0.6 MPa. Here the jet was formed within a De Laval nozzle with diameter of throat of 3 mm and divergence angle 1°. These experimental data clearly show that for a dilute jet ($\epsilon < 3 \times 10^{-4}$) the particle velocity is independent of jet concentration at a given gas flow rate (Fig. 6.2(a)). Above a solid volume fraction of 3×10^{-4}, however, particle–particle interaction can no longer be neglected and the velocity may be seen to decrease. It is obvious that this concentration threshold depends on particle size and, as shown in Fig. 6.2(b), shifts to lower concentrations for smaller particles. One can see that the extrapolation of power approximations up to lower concentrations leads to the reasonably accurate prediction of particle velocities. Thus, simulations of heavily laden

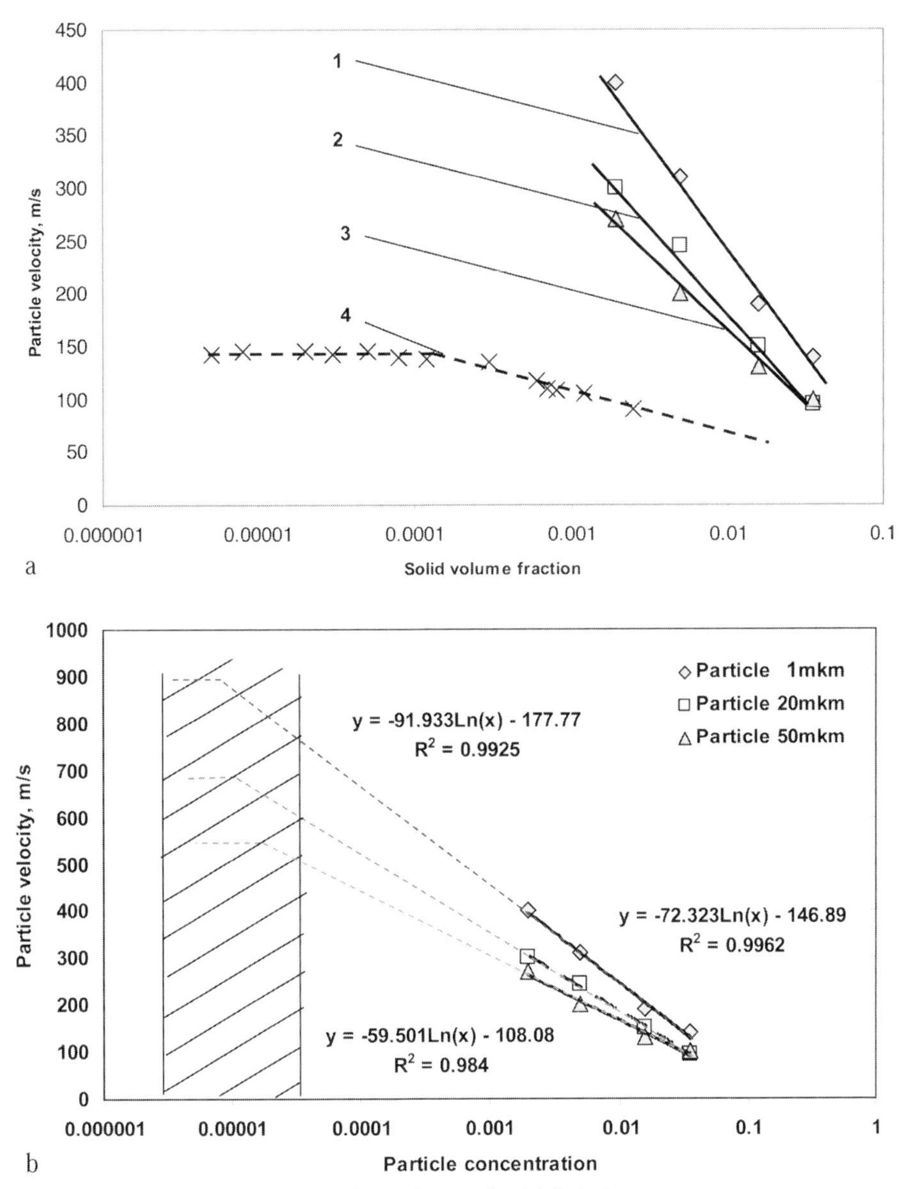

Fig. 6.2 Effect of particle concentration in the nozzle. (a) Calculation data; (b) power approximations. (1) modeling particle 1 μm; (2) modeling particle 20 μm; (3) modeling particle 50 μm; (4) experimental results. (1), (2), and (3) – after Eskin et al. (1999); (4) air 0.6 MPa, particle 120 μm after Mebtoul et al. (1996).

gas-particle jets using an engineering model of particulate friction – as developed by Eskin and Voropaev (2004) – permits the computation of gas-particle jet velocity parameters for a wide range of particle concentrations.

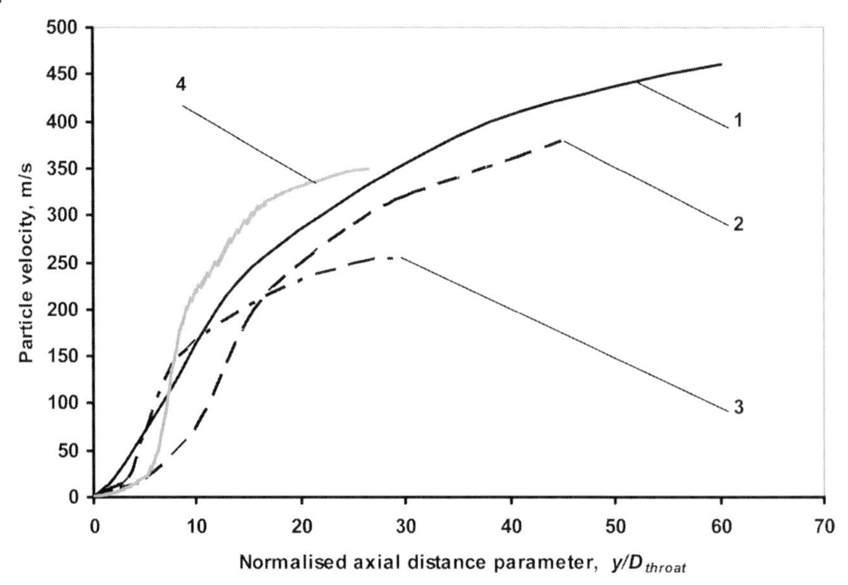

Fig. 6.3 Numerical simulation results by various authors: (1) Sakaki and Shimizi (2001); (2) Van Steenkiste et al. (1999); (3) Jen et al. (2005); (4) Eskin et al. (1999). In most of these studies copper powder with particle size of 50 μm was used; Eskin et al. (1999) used cement powder with density 3 g/cm^3; Nitrogen (Sakaki and Shimizi (2001), Jen et al. 2005) and air (Van Steenkiste et al. (1999), Eskin et al. 1999) were used as carrier gases with inlet pressures of 2.0 MPa and 0.6 MPa and temperatures of 800 K, 750 K, and 350 K. The data of Eskin et al. (1999) were extrapolated to a solid volume concentration about 1×10^{-5}.

Various one-dimensional simulations of gas-particle jets and the dependency of particle velocity on the normalized axial distance parameter (x/D_{throat}) have been undertaken over the years. A comparison of these works by various authors is shown in Fig. 6.3. Analysis of these results show that the numerical simulations of Van Steenkiste (1999) and Sakaki and Shimizi (2001) yield higher values of particle velocity than the more detailed simulation procedure developed by Jen et al. (2005). The method presented by Jen et al. (2005) also allows the analysis of some more complex phenomena such as shock waves, flow separation, the effects of wall friction, etc. As a result, the axial particle velocity at the nozzle exit is affected by frictional forces within divergent section of the nozzle, causing the velocity of carrier gas to decrease considerably. The results of Eskin et al. (1999) yield results similar to Sakaki and Shimizi (2001) for axial particle exit velocities. This agreement suggests that it is indeed possible to approximate particle behavior within the gas-powder jet using an "engineering model."

6.3.2 Influence of Gas Pressure

Previous experimental results have demonstrated the relationship between the variation of the gas inlet pressure and the velocity of the particles at the exit of the nozzle, mass flow rate of the gas, the solid/gas mass flow ratio, and other parameters of the gas-powder jet. As suggested by Jodoin et al. (2006), however, it is the nozzle mass flow rate that should be chosen as a global examination parameter because:

i. It is not an input variable of the model but rather a result from it;
ii. It can be monitored precisely within the experimental setup;
iii. It contains all the relevant physics of the process.

In fact, the mass flow rate may be predicted through one-dimensional isentropic analysis by means of the relationship,

$$\dot{m} = \frac{p_0}{\sqrt{T_0}} A^* \sqrt{\frac{\gamma}{R}} \left(1 + \frac{\gamma - 1}{2}\right)^{\left(\frac{\gamma-1}{2-2\gamma}\right)} \tag{6.32}$$

The experimental results of Jodoin et al. (2005) and curves calculated by Eq. (2.39) are shown in Fig. 6.4 for nitrogen and helium at a temperature of 750 K. In both the cases, due to the complex nature of gas-powder flow inside the nozzle (including friction and shock wave effects), Eq. (6.32) permits only an approximation of the mass flow rate values. In fact, the experimental data reveal that the mass flow rate dependence is particularly nonlinear for relatively low stagnation pressures (up to 1 MPa). This appears to stand in contradiction with Eq. (6.32), where the mass flow rate scales proportionally with the stagnation pressure. Certainly this issue requires further study.

Results of modeling the particle velocity dependence on gas pressure are presented in Fig. 6.5. Also shown in this figure are the experimental data for a wide range of gas pressures (for nitrogen), as generated by Sakaki and Shimizi (2001). There are two areas of gas pressure within which the axial particle velocity dependences differ considerably. Also, for relatively low gas pressures, the difference between gas and particle velocity is lower than for high gas pressures. This may be attributed to the drag coefficient dependence on the pressure, as was shown in previous analysis (see Eq. (6.16)). The particle velocity calculation of Eq. (6.17) (in accordance with Alkhimov et al. 2001) clearly shows that at high gas pressure there is only a small discrepancy between the modeling results of Sakaki and Shimizi (2001) and Alkhimow et al. (2001) (see Fig. 6.5). Supersonic flow theory predicts that particle velocity should vary as the log of the stagnation pressure (Smith and Dykhuizen 1988, Dykhuizen and Smith 1998). Figure 6.5 show the experimental data of Gilmore et al. (2005) which are well fit by a logarithmic curve for high gas

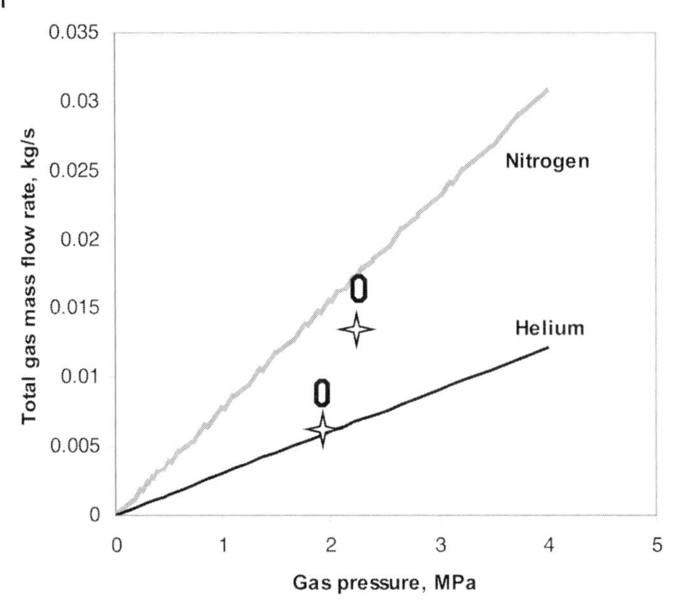

Fig. 6.4 Pressure dependence of mass flow rate after Jodoin et al. (2006). The curves are calculated by Eq. (6.32). Circles (nitrogen) and stars (helium) depict the experimental results.

pressures. However, the dependences of particle velocity on the gas pressure for relatively low pressures have not yet been studied in detail.

The main conclusion that may be drawn from the above discussion is that variations in gas pressure up to 1 MPa result in a great increase in particle velocity. Thus, there appears to be considerable opportunity in this region for particles to reach velocities that approach 500–550 m/s. Of course, the maximum velocity that may be attained is ultimately dependent on particle size and density.

6.3.3 Effects of Particle Concentration

An effect of jet particle concentration on the process characteristics of high-pressure GDS has been studied by Gilmore et al. (1999) and Van Steenkiste (1999) The results of this modeling suggest that above the mass powder/gas flow ratio of 0.03%, the mean particle velocity decreases linearly with an increase in the powder feeding rate. Gilmore et al. (1999) believe that this decrease in particle velocity might be due to the increased mass that must be accelerated by the gas flow. Unfortunately, this effect cannot be examined in detail because the increase in powder feeding rate leads to an overburdening of the nozzle with powder (Van Steenkiste et al. 1999). Nevertheless, the particle concentration greatly influences particles interaction, and in turn, the ac-

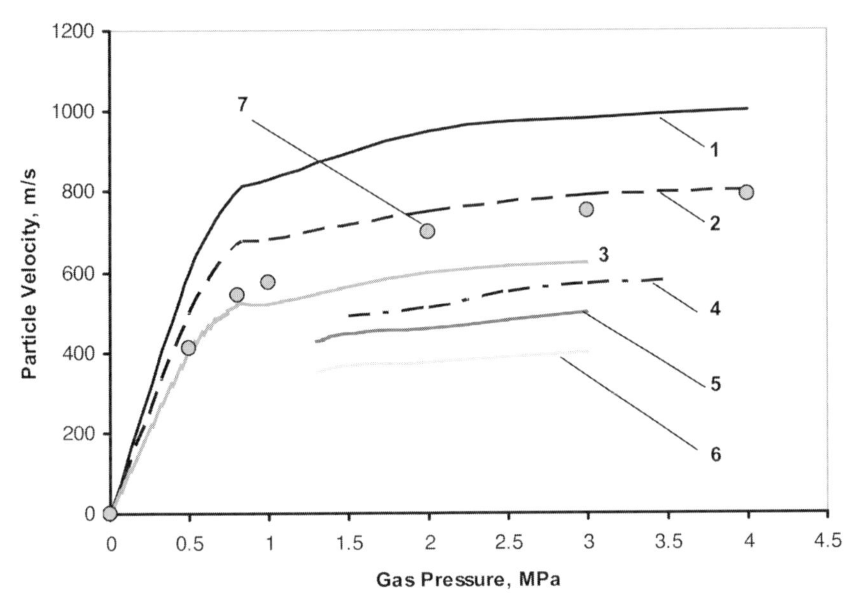

Fig. 6.5 Modeling of gas pressure influence. (1) Nitrogen, 750 K (Sakkaki and Shimizi 2001); (2) bronze particles 10 mkm, 750 K (Sakkaki and Shimizi 2001), (3) bronze particles 20 mkm, 750 K (Sakkaki and Shimizi 2001); (4) copper particle 15 mkm, 593 K (Stoltenhoff et al. 2002); (5) copper particle 22 mkm, Air 573 K (Gilmore et al. (2005); (6) copper particle 22 mkm, air 298 K (Gilmore et al. (2005), (7) calculation with Eq. (6.17), bronze particles 10 mkm.

celeration process. Controlling the powder concentration becomes even more important when powder is radially injected directly into the divergent portion of nozzle (Fig. 5.2(a)). In this case heavily laden particle jets may be formed.

Some calculations regarding an Al powder-laden jet within radial injection low-pressure GDS (Fig. 6.6) show that variations in the powder feeding rate strongly affect average interparticle distances. In fact, increasing the powder feeding rate M_S from 0.1 g/s to 6 g/s results in interparticle distances which range from 2.12 mm to 0.035 mm. As a result, the role of interparticle and particle-wall collisions is enhanced by many times. Particle concentration also influences particle behavior during formation of the powder layer. Experimental results for the determination of the GDS deposition rate for the case of radial injection low pressure are shown in Fig. 6.7. Here it can be seen that an increase in particle volume concentration leads to optimal results, increasing the build up capacity (thickness per pass) by five or more times. It is suggested by the authors that it is the collective behavior of particles which greatly influences coating formation during GDS. This postulate will be examined in the following sections.

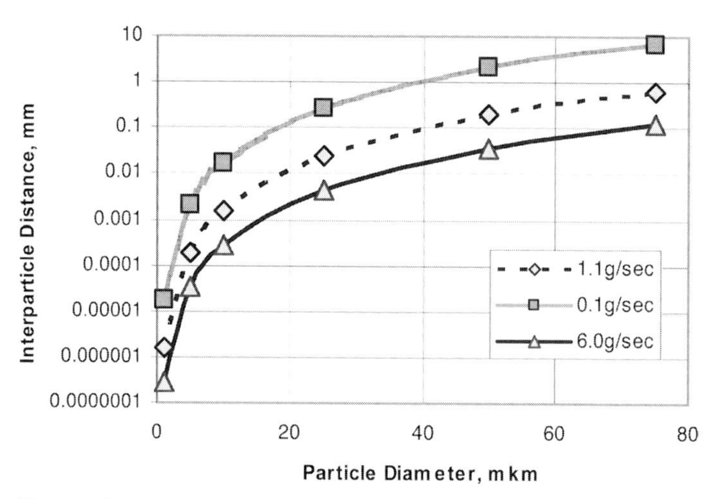

Fig. 6.6 An interparticle distance in an Al powder jet. $D_{\mathrm{throat}} = 3.5$ mm, air consumption, 0.022 m^3/s = 1.3 m^3/min.

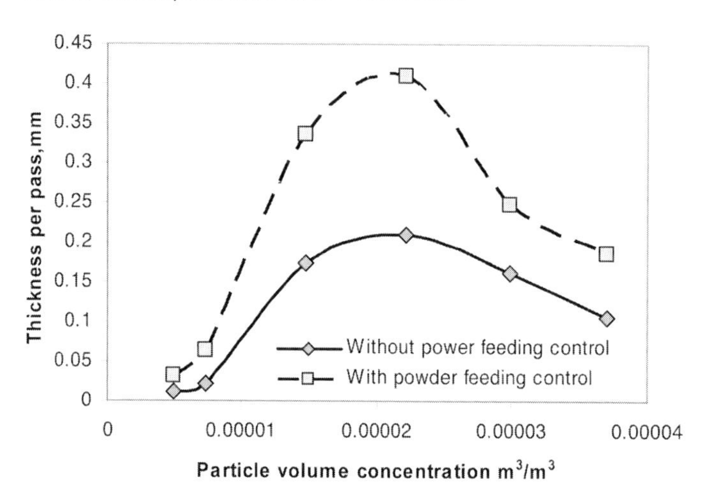

Fig. 6.7 Build up effectiveness of the radial injection low-pressure GDS of Al–Zn–Al_2O_3 powder mixture. Air pressure, 6 bar; air temperature, 300°C; air flow rate, 0.022 m^3/s; stand-off-distance, 10 mm; travel speed 10 mm/s.

6.3.4
Effects of Nozzle Wall Friction

As described previously, particle accelerations and particle velocities are generally calculated on the basis of isentropic flow using one-dimensional numerical particle acceleration models. The effects of wall friction, however, have

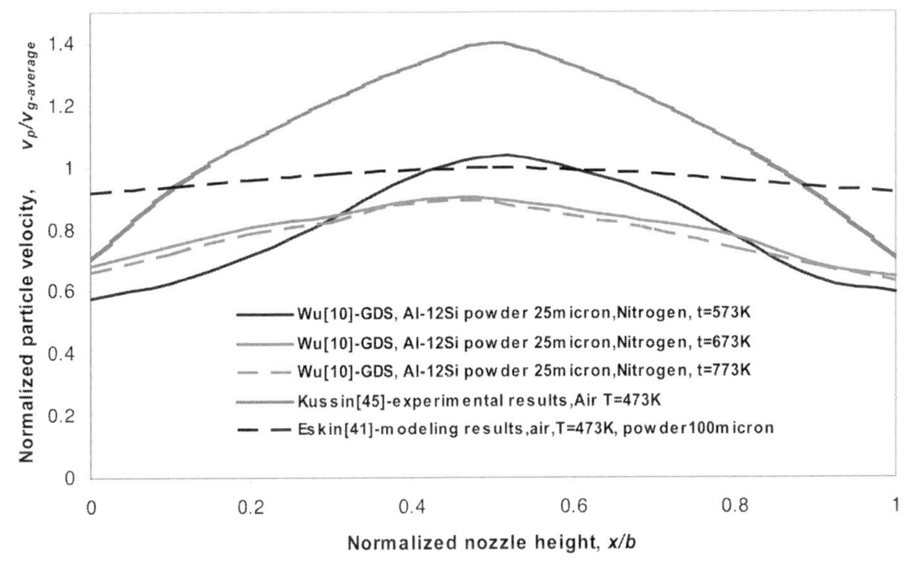

Fig. 6.8 Effect of nozzle wall friction on the gas and particle velocity distribution: Experimental (Wu et al. 2005, Kussin and Sommerfeld 2002) and modeling (Eskin 2005) results comparison.

not yet been explored in detail. For this reason there is little experimental data available regarding the velocity distribution of in-flight particles within the GDS process (Wu et al. 2005). Nevertheless, this problem is being investigated deeply using theories of multiphase flow in both numerical and physical modeling. The effect of nozzle walls on a high-speed, two-phase gas particle flow can be defined by interparticle and particle-wall collisions, changes in particle concentrations in the near-wall region, axial particle velocities, etc. In order to compare dissimilar experimental conditions directly, these dependences must be characterized in normalized, nondimensional coordinates. To do this, x/b is considered to be the normalized channel height; the particle velocity v_p is normalized through division by the average gas velocity, i.e., $v_p/v_{g-\text{average}}$. A distribution of this velocity parameter $v_p/v_{g-\text{average}}$ through the nozzle channel up exit from the nozzle (calculated via particle velocity measurements by Wu et al. 2005) is shown in Fig. 6.8. Here they are compared with the modeling results of Volkov et al. (2005). The experimental data of the air velocity distribution presented by Kussin and Sommerfeld (2002) are also shown in this figure for illustration purposes.

The profiles of the particle velocities in comparison with the single-phase air velocity show that an increase in particle size results in a decrease in the particle velocity gradient. It can be seen in Fig. 6.8 that in areas near the nozzle wall the particles travel at velocities which exceed the air flow due to transverse dispersion and the slip boundary conditions. The increase in particle-size diminishes the response of the particles to the gas flow, in turn enhancing transverse dispersion due to wall collisions. As a result, the streamline particle velocity profiles exhibit less dramatic variation. The experimental results of Wu et al. (2005) reveal that the effect of gas temperature on the profiles of the particles velocities is similar to those observed with the change in particle size (Fig. 6.8). Perhaps this is because the increase in gas temperature causes a subsequent increase in the frequency of wall collisions. This, in turn, increases momentum losses due to particle-wall collisions and, therefore, the particle velocity distribution through the nozzle becomes more uniform. Hence, the particle size distribution, particle concentration and gas-powder jet temperature are the main parameters which influence the velocity profile and the loss of kinetic energy of the particles.

The longitudinal distribution of the radial particle velocity parameter and the specific frequency of particle-wall collisions are shown in Fig. 6.9 for a nozzle with $D_{\text{throat}} = 17$ mm (Eskin et al. 2003). As can be seen, an increase in particle size results in a decrease in the frequency of particle-wall collisions, as well as the radial particle velocity. It is perhaps obvious that the loss of kinetic energy should be considerably higher for the nozzles used in LPGDS with $D_{\text{throat}} = 2.5$–3 mm. This is because particles in these nozzles need travel a shorter distance to collide against the nozzle wall. In fact, the thickness of this "colliding layer" h_i (Eq.(6.29)) is dependent on the particle velocity. The estimation of the "colliding layer" thickness h_i for a typical GDS nozzle reveals a values of $h_i = 0.08$–0.2 mm. Thus, the nozzle wall friction during LPGDS results in considerable losses in particle kinetic energy within the nozzle wall area. This diminishes the deposition efficiency of LPGDS processes because energy losses in the nozzle wall area permit the particles from achieving the critical velocity threshold. A numerical simulation of gas-particle flow inside a straight cylindrical nozzle has been conducted by Jen et al. (2005). Results of this study indicate that the friction in this section results in a significant decrease in both gas and particle velocity, confirming the assumption presented above.

6.4 Free Jet Characterization

6.4.1 Shock Wave Features of the Jet

In the research discussed above, the analysis of gas-particle flow was focused on supersonic flow inside the nozzle. As mentioned, the most important parameters having influence on the dynamics of the jet gas are temperature, par-

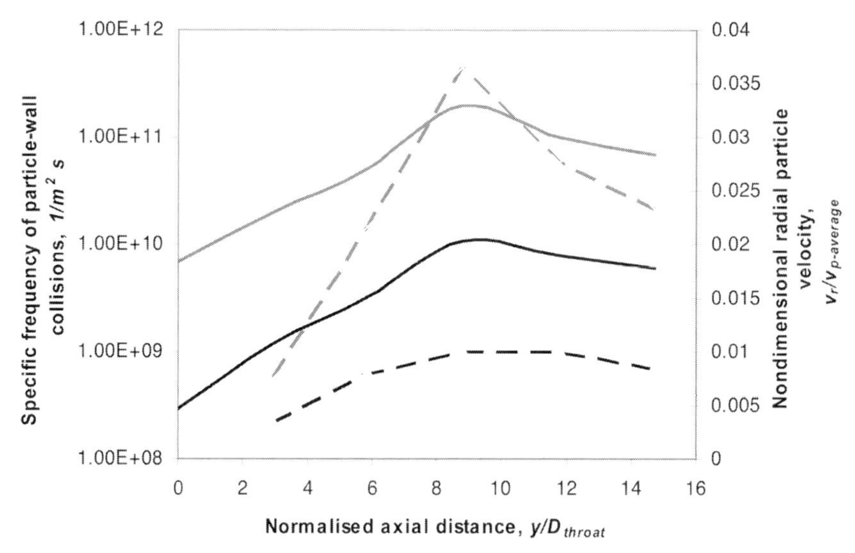

Fig. 6.9 Axial distribution of the radial particle velocity parameter and specific frequency of particle-against-wall collisions, after Eskin et al. (2003).

ticle velocity, and the concentration of particles in the powder-laden jet. In actuality, the kinematics of particle movement is strongly affected by particle flow outside the nozzle as well (Fig. 5.2(b)). As such, the simulation of gas flow before and after exiting the nozzle has been completed by Jen et al. (2005). This data, shown in Fig. 6.10, reveals that compression waves occur both inside (Fig. 6.10(a)) and outside (Fig. 6.10(b)) the nozzle. Moreover, the multiple reflections of the shock waves result in velocity fluctuations. These shock fluctuations are formed due to the pressure difference between the gas jet and the ambient atmosphere at the nozzle exit (Fig. 6.11). The densely spaced contour levels that can be seen in Figs. 6.10(b) and 6.11 – occurring outside the jet core – indicate the shear layers developing at the jet boundary. The relative thickness of these layers, $h_{\text{shear}}/D_{\text{jet}}$, is about 0.1–0.15, where h_{shear} is the thickness of layer in which the particle velocity falls from 0.5 v_m to 0. It should be noted that these values exceed the thickness of the colliding layer within the nozzle (Eq. (6.29)).

As previously shown, the calculated values of h_i for a GDS nozzle with $D_{\text{throat}} = 3.0$ mm are in the range of 0.08–0.2 mm. Thus, within the nozzle the ratio h_i/D_i is found to be approximately 0.03–0.07. This suggests that it is the nozzle wall friction that leads to particle flow localization. However, the localization effect seems not to be visible when measuring particle velocity. For this reason the particle velocity results shown in Fig. 6.8 reveal smooth curves without sharp peaks.

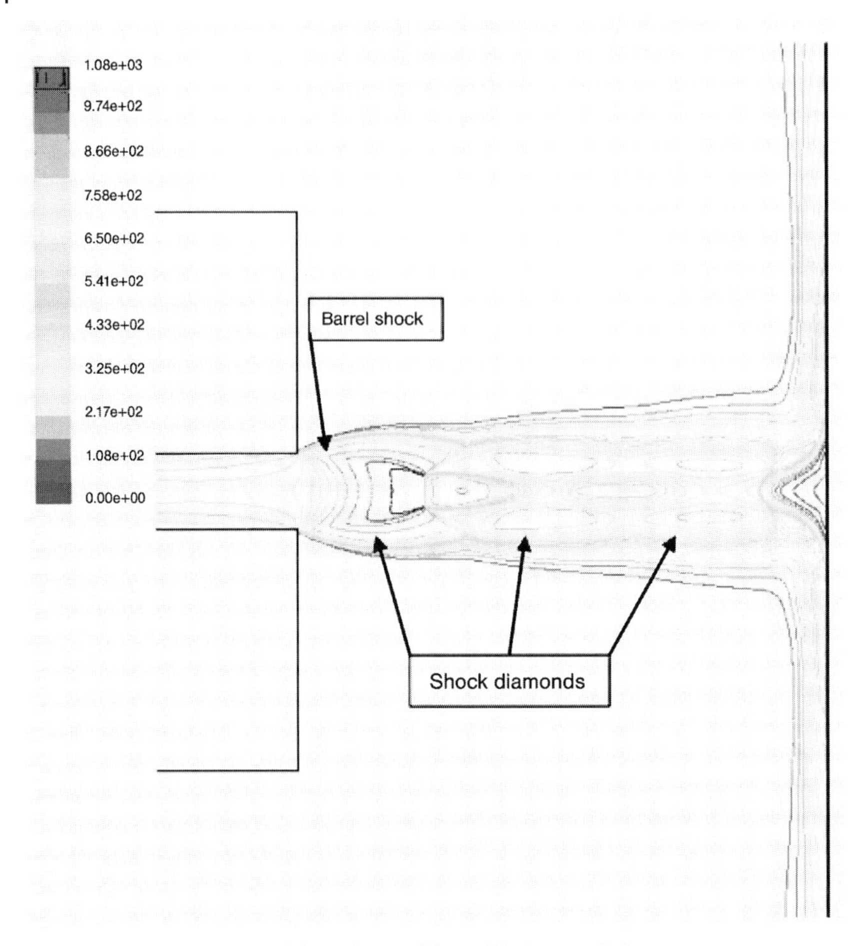

Fig. 6.10 Velocity contours (a) inside and (b) outside the nozzle for nitrogen carrier gas, after Jen et al. (2005).

Jen et al. (2005) have analyzed the principle features of the free jet structure. They state that there are three primary zones between the nozzle exit and the substrate: (i) the Mach disk – located close to the nozzle exit, (ii) a second weaker Mach disc, and (iii) a Bow shock formed on the substrate (Fig. 6.11). One can see that the pressure in the area enclosed by the Bow shock and the substrate is much higher and vortex formation is observed. The existence of this Bow shock precludes small particles (< 0.5 μm) from bonding to substrate.

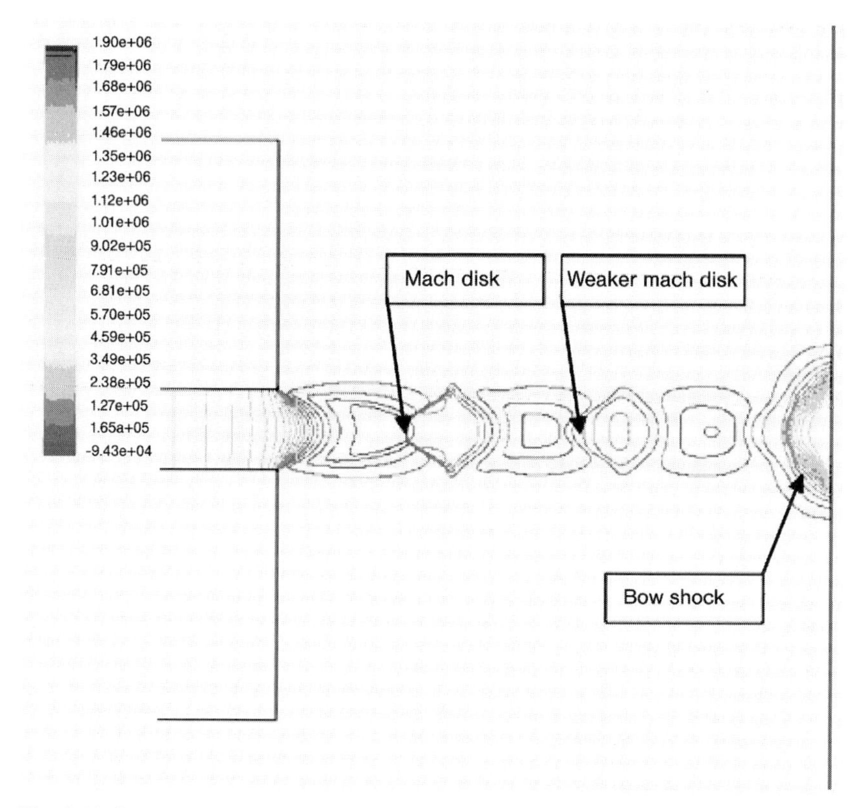

Fig. 6.11 Pressure contour outside the nozzle, after Jen et al. (2005).

6.4.2
An Engineering Model of the Free Jet

Numerical simulation is a powerful tool for studying GDS processes both inside and outside of nozzle. There are, however, significant influences on the particle-laden gas jet that may be characterized by simple relationships that are based on an engineering model of the free jet. This description of the nozzle free jet was first introduced by Kosarev et al. (2003); though relatively simple, it has proven quite useful in the calculation and analysis of real GDS processes. In particular, this method of analysis is based on the assumption that the profiles of velocity and dynamic pressure (ρv^2) are self-similar throughout the three regions of jet flow between the nozzle and the substrate.

In order to estimate the relationship between the velocity and temperature difference profiles, a normalized jet thickness parameter must be introduced: x/δ_M. Here δ_M is a jet thickness along the x-axis, which is defined as the distance from the jet axis to the point where $M^2(\delta_M) = 0.5M_m^2$. The normalized velocity parameter is $\phi_M = (M/M_m)^2$, where M_m is the axial Mach number.

Using this parameter the data obtained by Kosarev et al. (2003) may be fit by the function

$$\phi_M = \exp[-(0.83x/\delta_M)^2] \tag{6.33}$$

A similar equation can be used to approximate the stagnation temperature (T_0) at each point of the free jet.

It is commonly accepted in jet theory that the profiles of the stagnation temperature difference ($\Delta T_0 = T_0 - T_a$) are self-similar to the velocity profile. This suggests the relationship

$$(T_0 - T_a)/(T_{0m} - T_a) = \Delta T_0/\Delta T_{0m} = (v_g/v_m)^\sigma \tag{6.34}$$

where T_a is the ambient temperature, T_{0m} is the axial stagnation temperature, and σ is a ratio between the velocity profile thickness (δ_v) and the temperature profile thickness (δ_T):

$$\delta_v^2 = \sigma\delta_T^2 \tag{6.35}$$

More specifically, δ_T is the temperature profile thickness along the x-axis, which is defined as the distance from the jet axis to the point where $\Delta T_0(\delta_T) = 0.5\Delta T_{0m}$.

The experimental results shown in Fig. 6.12 reveal the similarity of the Mach and stagnation temperature profiles, allowing the gas temperature and velocity distribution to be described by Eq. (6.13). The ratio σ must be specifically defined for individual nozzle geometries; the type (axial or radial) and point of particle injection into the nozzle also seem to influence σ.

A y-axis (longitudinal) distribution of axial jet characteristics M_m and ΔT_{0m} are shown in Fig. 6.13 for the normalized parameters $(M_m/M_*)^2$ and $\Delta T_{0m}/\Delta T_{0^*}$, where M_* and ΔT_{0^*} are the parameters at the nozzle exit. The experimental data may be fit to the curve

$$\left(\frac{M_m}{M_*}\right)^2 = [1 + 3(y/y_{0.5}^M)^4]^{-0.5} \tag{6.36}$$

where y is the longitudinal coordinate in the jet and $y_{0.5}^M$ is the coordinate where $M_m^2(y_{0.5}^M) = 0.5M_*^2$. The approximation of the longitudinal distribution of ΔT_{0m} (Fig. 6.13) is consistent with the equation

$$\Delta T_{0m} = \Delta T_{0^*}/[1 + 2^8 - 1]\bar{y}^4]^{1/8} \tag{6.37}$$

where

$$\bar{y} = y/y_{0.5}^T, \; y_{0.5} \approx 2y_{0.5}^M \tag{6.38}$$

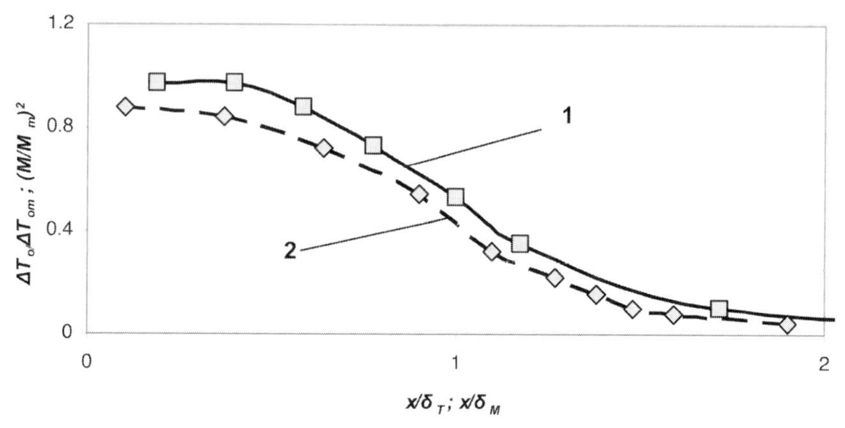

Fig. 6.12 A distribution of the normalized stagnation temperature difference and normalized M^2 in the free jet cross section. (1) $\Delta T_o/\Delta T_{om} = f(x/\delta_T)$; (2) $(M/M_m)^2 = f(x/\delta_M)$, after Kosarev et al. (2003).

Further, Kosarev et al. (2003) show that the relationship between the axial stagnation difference and the axial value of M^2 should be close to the form

$$\Delta T_{0m}/\Delta T_{0*} = (M_m^2/M_*^2)^{0.25} \tag{6.39}$$

One of the fundamental problems of jet theory is the determination of the jet thickness, or for a symmetric jet, the jet diameter as a function of the longitudinal coordinate. As stated by Abramovich (1963), a linear increase in thickness is observed both in the initial and primary regions of the jet, though with different proportionality coefficients. For the case of a plane jet (i.e., one in which its expansion is ignored in the direction perpendicular to flow), Kosarev et al. (2003) approximates this thickness with a nonlinear function

$$\delta_x = h[1 + 3(y/y_{0.5}^M)^4]^{-0.5} \tag{6.40}$$

Experimental data, on the other hand, are more accurately fit with (Fig. 6.14):

$$\delta_x = 0.75h[1 + 3(y/y_{0.5}^M)^4]^{-0.4} \tag{6.41}$$

Therefore, the semiempirical equations (6.33)–(6.41) permit characterization of the gas flow parameters of the free jet. Consequently, this facilitates a means by which it is possible to define their influence on the structure formation process.

6.4.3
Particle Flow Structure Within the Normal Shock Region

The effect of high-speed, two-phase gas-particle jet impingement on a solid body is of significant interest in many applications such as aircraft and turbine

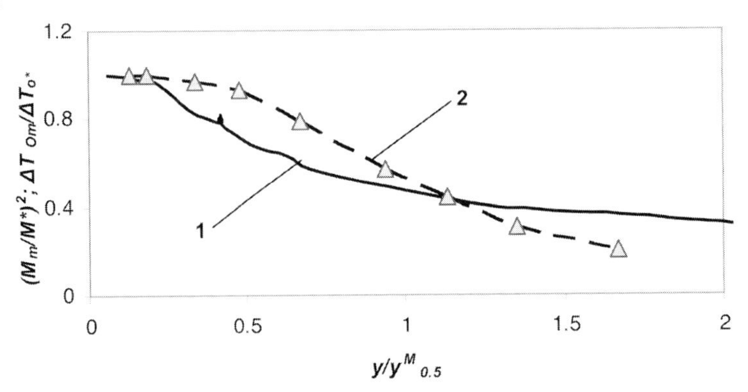

Fig. 6.13 A distribution of the normalized stagnation temperature difference and normalized M^2 versus the longitudinal coordinate. (1) $1 - \Delta T_{om}/\Delta T_o^* = f(y/y_{0.5})$; (2) $(M_m/M^*)^2 = f(y/y_{0.5})$, after Kosarev et al (2003).

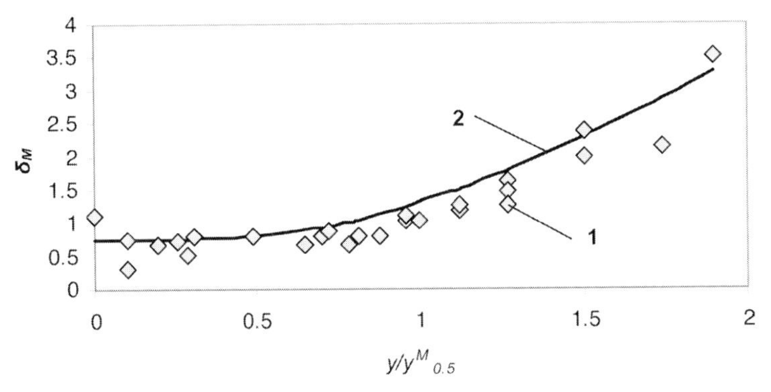

Fig. 6.14 A dependence of jet thickness versus the longitudinal coordinate. (1) Experimental results; (2) Eq. (6.41), after Kosarev et al. (2003).

design, industrial processing, etc. The presence of particles in the free stream – even if their concentration is very low – modifies the flow properties as compared to those of a pure gas (Volkov et al. 2005). The mechanics of aerosols, primarily developed by Fuchs (1997), suggest that for given body shape and flow parameters, there is a critical value of particle radius r_{p^*} above which bonding will take place. However, despite exceeding this critical size, upon impingement, particles may either bounce off or stick to the body surface. If the particle concentration in the free jet is sufficient, the reflected particles can collide with the incident ones, resulting in the formation of a near-wall layer in which particles move chaotically, colliding with each other. Theoretical estimations show that for coarse-grained particles (the radii of which are much more than r_{p^*} collisions between incident and reflected particles during flow over a blunt body can play a noticeable role, particularly if particle volume fractions are minimal (at $\alpha_{p\infty} \sim 10^{-6}$, Tsirkunov 2001).

6.4.4 Particle Collisions

The importance of collisions between particles in two-phase gas-particle jets has been demonstrated in studies of flow in pipes and channels (Sommerfeld 2003), and also in the analysis of particle dispersion in homogeneous isotropic turbulence (Sommerfeld and Kussin 2003, Tanaka and Tsuji 1991, and Oesterle and Petitjean 1993). The effect of interparticle collisions on wall erosion and heat transfer by particles in an impinging particle-laden jet has been studied numerically with the use of the direct simulation Monte Carlo (DSMC) method by Kitron et al. (1981). Volkov et al. (2005) further clarify the role of interparticle collisions and two-way coupling effects in a dusty-gas flow over a blunt body, particularly inside the boundary layer. This paper provides detail descriptions of various models of gas and dispersed phase flows, including gas-particle interactions, particle–particle collisions and particle–wall impact interactions. The fine flow structure of each phase is studied with respect to the free stream particle volume fraction $\alpha_{p\infty}$ and the particle radius r_{p^*}.

The computational flow model of Volkov et al. (2005) represents a combination of a CFD method for the carrier gas and a Monte Carlo method for the "gas" of particles. The numerical results illustrate many important features which are commonly observed in high-speed two-phase flows over blunt bodies. Of particular interest in this work is the calculated particle flow structure in the area of particle impingement. The particle phase flow patterns were found to be quite different for fine particles ($r < r_{p^*}$) and coarse-grained particles($r > r_{p^*}$). Figure 6.15 shows the distribution of the volume fraction for particles which collide with a blunt body of radius R. The role of interparticle collisions is shown to be negligible for fine particles (< 1 μm); whereas, the effect of collisions on the behavior of larger particles is more significant as the particle content $\alpha_{p\infty}$ increases. This results in a sharp increase in the particle concentration near the surface of substrate. In fact, some particle concentration peaks are seen only for low particle concentrations (curve 1). The longitudinal distribution of $\alpha_p/\alpha_{p\infty}$ changes qualitatively when $\alpha_{p\infty}$ increases from 10^{-6} to 3×10^{-5}. Unfortunately, numerical simulation data are available only for particles of size < 1 μm. Nonetheless, these results demonstrate the strong longitudinal dependence of the concentration profile at the near-surface field.

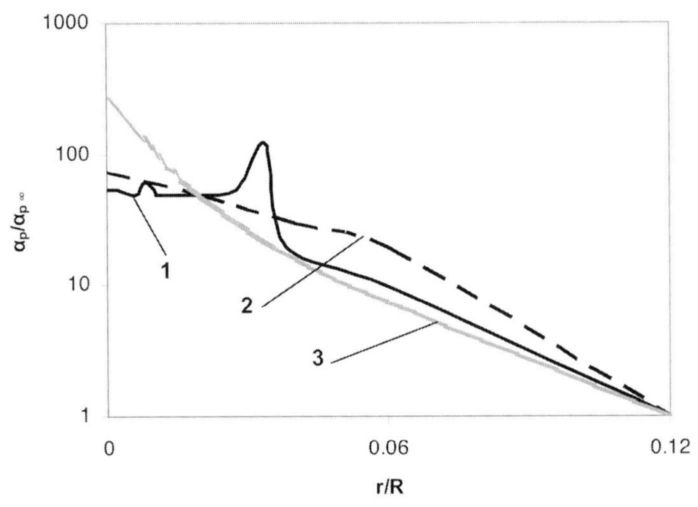

Fig. 6.15 Distribution of a relative particle volume fraction $\acute{\alpha}_p = \alpha_p/\alpha_{p\infty}$ for particles with $r_p = 1$ µm: (1) $\alpha_{p\infty} = 3 \times 10^{-6}$; (2) $\alpha_{p\infty} = 10^{-5}$; (3) $\alpha_{p\infty} = 3 \times 10^{-5}$.

6.5 Concluding Remarks

It can be concluded that the particle acceleration process in powder-laden jets during gas dynamic spray depends on three groups of factors: nozzle structure, gas and parameters, and powder characteristics. The most significant of these are nozzle geometry, type of carrier gas, pressure and gas temperature, particle size and density, particle concentration in the powder-laden jet, and friction effects. The LPGDS process is characterized by specific conditions of interparticle and particle – nozzle wall friction. This is due to the high particle concentration of the powder-laden jet. Engineering models and numerical simulation of particle acceleration permit evaluation of the main process parameters by taking into account the friction and Bow shock effects. As an example, evaluation of the deposition efficiency and shock wave effects during GDS may be accomplished by applying the results of isentropic numerical modeling, as presented in Chapter 7.

7 Deposition Efficiency and Shock Wave Effects at GDS

The GDS system accelerates micron-sized particles to high velocities by entraining the particles in the flow of a supersonic nozzle (Fig. 1.1). In general, as previously shown, high velocities are necessary to optimize particle deposition and coating density. Other parameters such as gas conditions, particle characteristics, and nozzle geometry also affect the particle velocity and deposition efficiency. This section illustrates how these parameters affect the coating characteristics, and is based on the work of Helfritch and Champagne (2006).

7.1 Model Structure

7.1.1 Statement of Task

The modeling of deposition efficiency, or particles building up over previously deposited particles, can be broken down into three tasks:

1. The gas flow and temperature in the nozzle are calculated by means of isentropic (frictionless) gas dynamic principles.
2. Drag and heat transfer coefficients from solid rocket analyses are used to iteratively calculate particle velocity and temperature through the nozzle.
3. An empirical relationship between the particle velocity and particle material characteristics is used to calculate the deposition efficiency or the percent of incoming particles that adhere and form the coating.

7.1.2 Gas Flow

The gas flow model uses isentropic relationships and linear nozzle geometry. The assumptions for the calculation are as follows:

Introduction to Low Pressure Gas Dynamic Spray Roman Gr. Maev and Volf Leshchynsky

ISBN: 978-3-527-40659-3

- The gas obeys the ideal gas law.
- There is no friction impeding the gas flow.
- The gas flow is adiabatic, i.e., no heat exchange occurs with the surroundings.
- Steady-state conditions exist.
- Expansion of the gas occurs in a uniform manner without shock or discontinuities.
- Flow through the nozzle is one-dimensional; hence, the flow velocity, pressure, and density are uniform across any cross section normal to the nozzle axis.
- Particles do not influence gas conditions.

Under these conditions, the Shapiro (1953) relationship between the nozzle area A and the Mach number is given by Eq. (7.1), where γ is the ratio of gas specific heats (C_p/C_v):

$$\frac{A_1}{A_2} = \frac{M_1}{M_2}\left\{\frac{1+[(\gamma-1)/2]M_1^2}{1+[(\gamma-1)/2]M_2^2}\right\}^{\frac{\gamma+1}{2(\gamma-1)}} \tag{7.1}$$

The simple conical nozzle geometry that is shown in Fig. 5.2(c) is assumed. A small initial subsonic Mach number and initial (stagnation) values of pressure and temperature are assigned at the converging section of the nozzle. The Mach number is then iteratively increased, while gas characteristics are calculated for each point through the isentropic relationships of Eqs. (7.1) and (7.3). Linear progression along the nozzle axis is calculated from the area change given by Eq. (2.49) and the assumed nozzle geometry.

$$\frac{T_0}{T} = 1 + \left(\frac{\gamma-1}{2}\right)M^2 \tag{7.2}$$

$$\frac{p_0}{p} = \left[1 + \left(\frac{\gamma-1}{2}\right)M^2\right]^{\frac{\gamma}{\gamma-1}} \tag{7.3}$$

The flow between the nozzle exit and the shock wave upstream of the substrate is assumed to be constant, and equal to the conditions at the nozzle exit. The stand-off position of the shock wave relative to the substrate is given by Billig's approximation (Billig 1967) as

$$\Delta = 0.143 d_e \left[\exp(3.24/M_e^2)\right] \tag{7.4}$$

where d_e equals the nozzle exit diameter, and M_e is the Mach number at the nozzle exit. The Mach number immediately behind the shock wave is given by the normal shock relationship:

$$M^2 = \frac{\left[M_e^2 + \frac{2}{\gamma-1}\right]}{\left[\left(\frac{2\gamma}{\gamma-1}\right)M_e^2 - 1\right]} \tag{7.5}$$

Downstream of the shock wave, the Mach number is assumed to linearly decrease from the value given by Eq. (2.53) to zero at the substrate surface.

7.1.3 Particle Motion

Once the gas conditions and velocity are characterized, the particle velocity may be iteratively calculated from the nozzle entrance to the substrate through the use of a solid rocket nozzle particle drag relationship (6.16). This relationship predicts the accelerating force on the particle to be

$$m\frac{dV_p}{dt} = C_D(\pi/8)\rho_g d^2 (V_g - V_p)^2 (6.16a)$$

where V_p and V_g are particle and gas velocities, m is the particle mass, ρ_g is the gas density, and d is the particle diameter. Carlson and Hoglund (1964) correct the simple Stokes drag law relationship for inertial, compressibility, and rarefaction effects through the empirical relationship of Eq. (2.54):

$$C_D = \frac{24}{R_e}\left[\frac{\left(1+0.15R_e^{0.687}\right)\left(1+e^{-0.427/M_p^{4.63}+3.0/R_e^{0.88}}\right)}{1+(M_p/R_e)(3.82+1.28e^{-1.25R_e/M_p})}\right] \tag{7.6}$$

where M_p is the Mach number of the gas-particle velocity difference and R_e is the gas-particle Reynolds number.

Particle temperature is subsequently calculated through the application of the gas-particle heat transfer relationship for forced convection, given by Eq. (7.7):

$$c_p\frac{dT_p}{dt} = (N_u k/d_p)(A_p/m)(T_g - T_p) \tag{7.7}$$

where c_p is the particle heat capacity, T_p, and T_g are the particle and gas temperatures, N_u is the Nusselt number, k is the gas conductivity, and A_p is the particle surface area.

Ranz and Marshall (1952) show that the Nusselt number for this situation is given by

$$N_u = 2.0 + R_e^{0.5}P_r^{0.33} \tag{7.8}$$

where P_r is the Prandtl number.

7.1.4 Deposition Efficiency

The ability to predict the deposition efficiency allows one to choose gas and particle parameters that will yield effective deposition efficiency and in turn reduce both the operating time and wasted powder. An empirical relationship between particle parameters and the critical velocity necessary for a particle to stick to a previously deposited layer, as well as the gas dynamic velocity model shown above, are used to predict the deposition efficiency. The empirical relationship for the particle critical velocity, V_c, is given by Assadi et al. (2003) as

$$V_c = 667 - 14\rho_p + 0.08T_m + 0.1\sigma_\mu - 0.4T_e \tag{7.9}$$

where ρ_p is the particle density in g/cm^3, T_m is particle melting point, °C, σ_μ is particle ultimate strength, MPa, and T_e is particle exit temperature, °C.

The critical velocity as determined by Eq. (7.9) and the particle velocities as a function of diameter then allow an identification of the particle diameters that are able to achieve this velocity. It will in fact be shown that the diameter of particles that exceed critical velocity are limited at both the small end and the large end of a range of diameters.

The powders employed in cold spray are not of uniform diameter, but may be characterized by a distribution of particle diameters with the distribution defined as normal. Deposition efficiency is calculated as the percent of particles having a smaller diameter than the largest particle diameter achieving the critical velocity minus the percent of particles having a smaller diameter than the smallest particle that can achieve the critical velocity. For a normal distribution, Eq. (7.10) defines the mass percent of particles having a smaller diameter than the particle diameter d_p, where MMD is the mass mean diameter of the feed stock powder and s is the geometric standard deviation of the distribution. Once the smallest and the largest particles that reach the critical velocity are determined, the percent of particles smaller than the smallest particle achieving the critical velocity is subtracted from the percent of particles smaller than the largest particle. The result of this subtraction yields the deposition efficiency:

$$\% = \left(\frac{100}{2}\right)\left[1 + \text{erf}\left(\frac{d_p - MMD}{\sigma\sqrt{2}}\right)\right]$$

7.2 Calculations and Discussion

Nitrogen gas is used for these calculations. The gas stagnation (chamber) pressure and temperature are 2.76 MPa and 673 K, unless otherwise noted. The initial particle temperature is 293 K. The nozzle length is 10 cm, its throat diameter is 2 mm, and its area ratio is 4.

Spherical particles are assumed. The mass ratio of particles to gas in typical cold spray operation is less than 0.05; therefore, it is assumed that the pres-

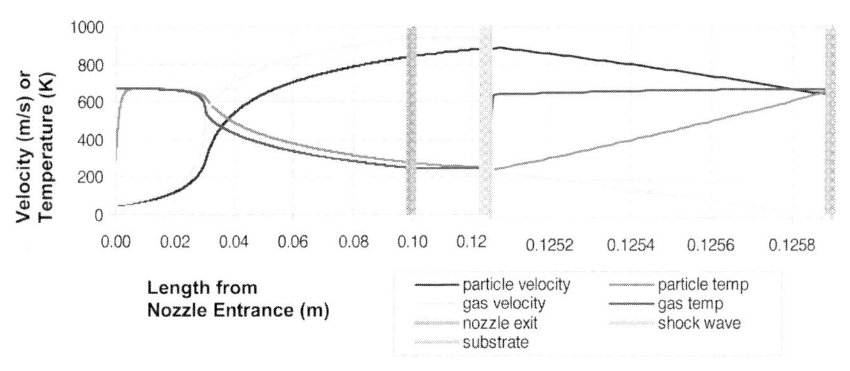

Fig. 7.1 Gas-particle nozzle flow impacting a substrate (note scale change).

ence of particles does not affect the gas flow and that no particle-to-particle contact occurs. Thus, an individual calculation applies to a single particle of a given diameter. Particle size effects may in turn be determined by multiple calculations for various particle diameters.

Figure 7.1 shows a gas-particle calculation for 3 μm copper particles. Nitrogen gas, initially at 2.76 MPa and 673 K, is the main gas. The gas converts temperature and pressure into velocity as it is expanded in the converging–diverging nozzle. The gas attains Mach 1 at the throat and is about Mach 3 at the nozzle exit. Particle velocity is related to the gas velocity through the drag relationship, Eq. (6.16a). It can be seen in this case that gas and particle exit velocities of 950 m/s and 850 m/s, respectively. Particle temperature is related to the gas temperature through convective heat transfer, Eq. (7.7). The particles are seen to heat up when they are cooler than the gas and begin to cool down after the gas has expanded to temperatures lower than that of the particles. The gas velocity is constant between the nozzle exit and the shock wave, while particle velocity increases somewhat.

The axis scale of the zone downstream of the shock wave is expanded in order to observe parameters in this length of less than 1 mm. As for all supersonic flow impacting an object, the flow correction is made by a shock wave which instantly reduces gas velocity to subsonic and increases both gas pressure and temperature. Particle velocity and temperature, however, are not instantly affected by the shock wave, but continue to be affected by gas velocity and temperature through drag and heat transfer, (6.16a) and (7.7). Gas velocity decreases linearly to zero downstream of the shock at the substrate surface. Accordingly, particle velocity also decreases and is at a value lower than that at the nozzle exit when it strikes the substrate. Thus, the use of nozzle exit velocities for deposition estimates can give rise to incorrectly calculated deposition efficiencies.

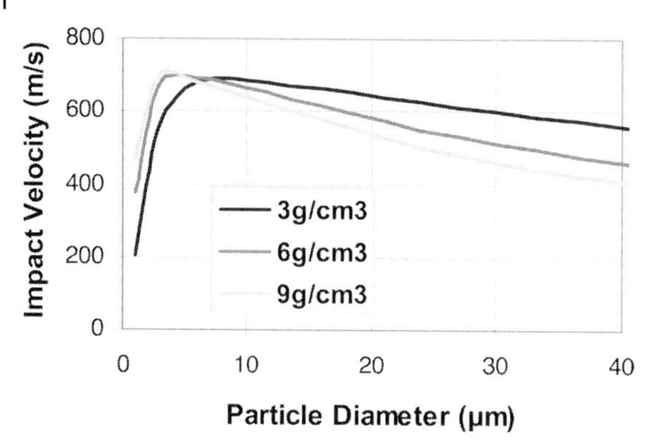

Fig. 7.2 Impact velocity as a function of the particle size for three particle densities.

Figure 7.2 shows how the particle impact velocity is affected by particle size for three values of particle density. This figure clearly shows that an optimum particle size exists for each density. Larger particles are relatively unaffected by the shock and achieve velocities consistent with the expectation that lighter (less dense) particles achieve higher velocities. The reverse is true for smaller particles, where lighter particles are decelerated most severely downstream of the shock wave. Large particles exit the nozzle at a lower velocity, but maintain this velocity past the shock wave. Therefore, there exists an optimum particle size which has a relatively high exit velocity but is relatively unaffected by the subsonic gas flow downstream of the shock.

Figure 7.3 results when the critical velocity for particle deposition (Eq.(7.9)) is included with the results of Fig. 7.2 for Copper particles ($\rho_p = 9\ \text{g/cm}^3$). Particles between 1.5 and 15 μm are seen to exceed the critical velocity and are therefore deposited. On the other hand, particles smaller than 1.5 μm and larger than 15 μm do not deposit. Thus for the distribution of particles within this feed stock powder, the deposition efficiency would be the weight percent of particles between 1.5 and 15 μm.

For comparison, the deposition efficiency is calculated for the case when a shock wave is present, as well as when no shock is present and particles impact at nozzle exit velocity. The deposition efficiency is calculated as a function of the mass mean diameter of Copper powder having a normal distribution and a standard deviation of 4. Figure 7.4 shows the results of these calculations, indicating that the presence of the shock wave has essentially no effect when spraying powders with distributions of mass mean diameter greater than 10 μm; however, shock-induced effects become large for smaller MMDs. This occurs because large particles are not affected by the reduced gas velocity during the short period of time between the shock and the substrate, while small particles are affected.

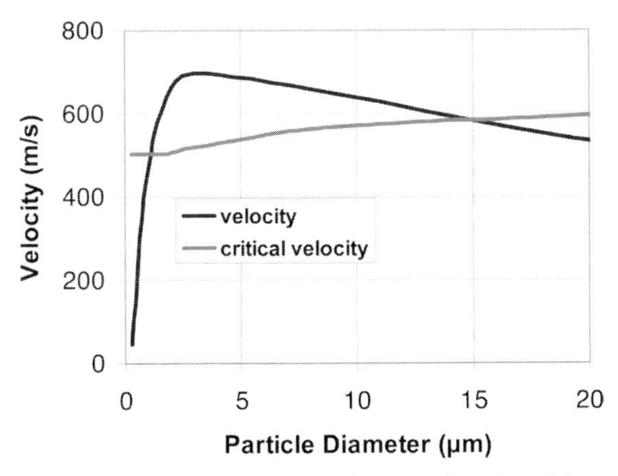

Fig. 7.3 Impact and critical velocities as a function of the particle size.

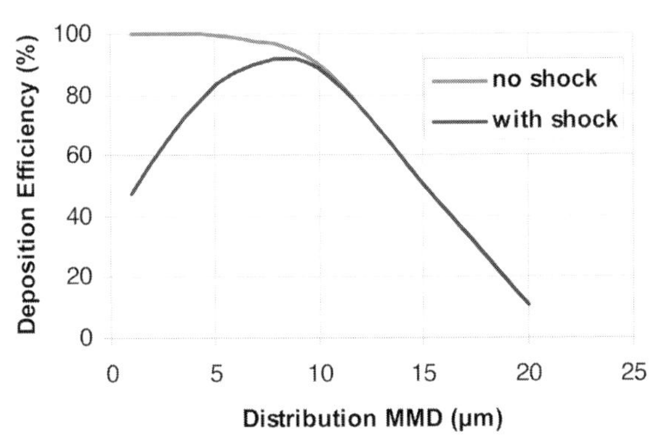

Fig. 7.4 Deposition efficiency as a function of the particle distribution mass mean diameter (MMD).

7.3 Critical Velocity Evaluation on the Basis of Rebound and Adhesion Phenomena

As discussed previously in Chapters 2 and 3, the particle rebound and adhesion phenomena may be defined by both elastic and plastic deformation of the particle and the substrate. Wu et al. (2006) show that during the impinging process, some of the initial kinetic energy of the in-flight particle is lost to the plastic deformation; the remaining energy is stored in the elastic zone during loading and is subsequently recovered during unloading. The rebound energy R is shown as

$$R = \frac{1}{2} e_r m_p v_p^2 \tag{7.10}$$

where e_r is the recoil coefficient for spherical particles which is given by

$$e_r = 11.47 \left(\frac{Y_{ef}}{E_{ef}}\right) \left(\frac{\rho v_p^2}{Y_o}\right) \tag{7.11}$$

Here Y_{ef} is the effective yield stress of the solid material, E_{ef} is the conventional elastic modulus, and ρ_p is the density of the particle material. Taking into account Eq. (7.11) the rebound energy (7.10) may be expressed as following:

$$R = 3.0 \left(\frac{Y_{ef}^{1/4}}{E_{ef}}\right) d_p^3 \rho_p^{3/4} v_p^{3/2} \tag{7.12}$$

In the Johnson–Cook equation (2.17) $Y_{ef} \approx s$ (Johnson and Cook 1983). On the other hand, the adhesion energy A of the particle impact is presented by Wu et al. (2006) as

$$A = \alpha_A A_{max} = \alpha_A \frac{\pi d_p^2 N_a E_1}{3(1 - \epsilon_f)} \tag{7.13}$$

where α_A is the fraction of bonded atoms per unit adhesive interface, and is also called the relative strength of the bond between the particle and the substrate. Also, N_a is the total number of atoms in the unit contact plane, E_1 is the energy of the single bond between two atoms, and ϵ_f is the total strain of the impinged particle. As it turns out, attachment of particle is achieved when $A \geq R$. Further evaluation by Wu et al. (2006) on the basis of Eqs. (7.12) and (7.13) reveals lower critical velocities (about 400 m/s), as presented in the Table 2.1. Moreover, by taking into account the considerably lower values of the apparent flow stress Y and the flow stress of solid material, Y_0, the real critical velocities of powder ensemble attachment appear to be below 400 m/s.

7.4 Concluding Remarks

Particle deposition and consolidation by GDS has been modeled through the application of conventional rocket nozzle flow equations and the application of an empirical materials-driven impact relationship. Particle velocities and temperatures were predicted at the nozzle exit, downstream of the bow shock and at the substrate surface. These conditions were subsequently used to predict the deposition efficiency.

It has often been thought that the smallest particles attain the highest velocities in the cold spray process, thus achieving good deposition efficiency. While it is true that small particles exit the nozzle at high velocity, their velocity at impact can be significantly lower. Modeling efforts showed that the low gas velocity following the bow shock wave decreases particle velocities, especially for the smallest particles. It was further shown that impact velocity increases as the particle diameter decreases until a diameter of 4 to 8 μm is reached. Impact velocity then decreases as the particle diameter is further reduced. This effect can decrease the deposition efficiency from 90% at a MMD of 8 μm to 50% at 2 μm. The modeling effort presented here provides a means to anticipate coating results based upon the spray parameters and material characteristics, thus eliminating trial and error attempts at creating acceptable coatings.

True values of critical velocity for GDS deposition of powder mixtures are considerably lower than for individual particles. It is for this reason the Low Pressure GDS (particle velocity is about 500–600 m/s) with high concentrations of powder material in powder-laden jet effective, and in fact preferable for numerous industrial applications.

8
Structure and Properties of GDS Sprayed Coatings

8.1
General Remarks

Over the past 25 years metal–matrix composites (MMC) have become key materials in the development of numerous technologies. In fact, composite coatings are used to create advanced high performance surfaces on mechanical components, die components, and sliding bearings, and biological implants. These applications often require high strength and/or toughness, among other use-specific properties. It is increasingly important to be able to adapt the properties of these materials to the various within which they are employed. In fact, because most MMC applications are structural in nature, it is essential to fully understand the factors which influence their mechanical behavior and, ultimately, their reliability. Therefore, accurate mechanical testing techniques are essential for the proper evaluation of both new and existing materials.

To characterize the mechanical properties of GDS coating materials, many authors have applied mechanical testing techniques such as hardness, tension, and bend tests Kreye (2004). However, a distinct lack of experimental results for GDS coatings and, in particular, for low-pressure GDS composite coatings permits deeper understanding of the structure formation process. Therefore, it is necessary to adequately compare experimental results with known structural properties in an effort to better understand and, in turn, optimize GDS technology

The primary microstructure parameters of powder composites are well known. These include porosity, pore size, matrix particle size, reinforcement particle size, volume fraction of the reinforcement particles, and interparticle distance between the reinforcement particles. Thus, the task of this chapter is to evaluate the mechanical properties of GDS coatings with respect to these microstructure characteristics. From this analysis, a novel approach to the systematic analysis of mechanical properties will be developed using ultrasonic and compression tests.

Introduction to Low Pressure Gas Dynamic Spray Roman Gr. Maev and Volf Leshchynsky

ISBN: 978-3-527-40659-3

8.2 Powder Materials for Low-Pressure Gas Dynamic Spray

8.2.1 Features of GDS Coatings

8.2.1.1 Microstructure

The severe plastic deformation of particles during GDS processes results in both the consolidation and strengthening of the resultant coating. This suggests that the extensive plastic flow of a particulate material is the main process which governs the structural formation of the coating. The metals are well-known materials which in many cases exhibit excellent ductility and low flow stress. Metal particles with high kinetic energy build up a dense coating due to deformation upon impact (Alkhimov et al. 1990). Thus, the macroscopic mechanical properties of composites are strongly conditioned by the bulk properties of the constituents; they are also influenced by the mechanical behavior of the matrix/reinforcement interfaces.

An overview of structure formation processes in GDS, as well as some microstructures, has been presented in Chapter 3. A detailed analysis of the bonding process has also been undertaken by Borchers et al. (2004). As shown in this study of the interfaces in high-pressure GDS process, interparticle bonding areas do not form a complete bonding network. In fact, some of the bonding areas exhibit microstructural features such as jets (as described above).

Using transmission electron microscopy (TEM), several remarkable features are noted at the interface by Borchers et al. (2004). In some locations within Al there appears a tangled structure, where the interface has bifurcations; here, a thin stripe of material with a thickness of some nanometers seems to be interlaced. The Al particles interface is in fact somewhat wavy on the order of some 100 nm, with the grains having low dislocation density. The grain size near the interface is about 500 nm. Copper coating GDS also exhibits a nonequilibrium grain boundary which is characterized by ultrahigh dislocation densities adjacent to the grain boundaries. The microstructural features Ni GDS coatings are similar to those of Cu. The particle–particle interfaces consist of nanosized grains and large grains with coffee-bean-like contrast, which is typical for shock consolidated structures (Meyers et al. 1999).

8.2.1.2 Interparticle Bonding

The above-described results clearly show that effective interparticle bonding takes place only in local areas. Moreover, the structure of the bonded areas is not uniform, influencing the interfacial mechanical properties as well. It seems reasonable to classify the interparticle bonds as those made by (i) the melting of adjacent particles, (ii) adiabatic shear band formation with a phase transformation at the interparticle contacts, and (iii) adhesion (Van Der Waals) forces.

At present there are two primary issues of GDS that have not yet been completely defined: (i) the type of interparticle bonds, and (ii) the surface area occupied by the definite type of interparticle bonds that are created by the appropriate powder consolidation technique. The primary bonding structures (described above) may be characterized by definite macroscopic mechanical properties. Thus, based on the analysis of the mechanical properties, the structure of interparticle bonds and the contact surface area data, it may in fact be possible to estimate the contribution of each bonding mechanism to the mechanical strength of the composite as a whole. Therefore, having characterized the interparticle bonding mechanism, it seems it will be possible to determine the (i) macroscopic bonding strength, (ii) bonding surface area, and (iii) bonding area structure.

An in-depth analysis of the interfacial mechanical properties of GDS sprayed materials has not yet been performed from this perspective. The most detailed information characterizing the bonding of Al, Cu, Ni, and Ti may be found in a series of papers by Borchers et al. (2003, 2004, and 2005) for HPGDS processes. These authors apply the Prummer (1987) classification of bonding mechanisms in samples prepared by explosive compaction. The authors uncover three main bonding types:

1. Adiabatic welding – the main bonding mechanism in the technique of explosive cladding. This bonding is characterized by adiabatic shear instability and jet formation at the welding interface. Prummer calls this bonding "explosive welding," but since cold gas spraying is not a technique where explosive are used, we use the above expression in this work.
2. Friction welding – here shear bands form which are not necessarily an adiabatic shear instability.
3. Liquid phase sintering – the near-surface areas of the powder particles melt during compaction and subsequently weld together during cooling. To find adiabatic welded interfaces, one has to look for jet formation as a result of adiabatic shear instabilities. Although it is not as easy to certify jet formation in LPGDS as it is in explosive welding, narrow bands of metallic materials observed between two particles could be evidence of jet occurrence (see Figs. 3.8 and 3.9(a)). Also, the plicate appearance of interparticle region on the SEM micrograph (Fig. 3.9(b)) may further suggest jet formation during LPGDS.

According to Borchers et al. (2005), the microstructure of HPGDS copper coatings is characterized by a narrow zone of dynamically recrystallized material around the particle–particle interfaces. Further, they find no evidence for microstructural phenomena related to melting. Despite the work of Maev et al. (2004, 2005, and 2006), there remains a decided lack of data concerning the microstructure characterization of composite coatings made by LPGDS. For this reason, the mechanical properties of the coatings created by HPGDS or LPGDS, as well as their relationship to the microstructure parameters have not yet been fully examined.

8.2.2
Overview of GDS Materials

A partial list of the materials successfully applied by HPGDS is shown in Table 8.1. These may be divided into three groups: (i) metals, (ii) alloys, and (iii) metal–matrix composites. Metal, ceramic and metal–matrix composite (MMC) coatings are finding increasing use in a wide range of applications for automotive, household, machinery, and electronic components. A large variety of MMCs are employed, but by far the most widely studied materials are Al metal–matrix composites. Metal–matrix composites that are reinforced with ceramic particles offer high strength and a high elastic modulus, as well as favorable high-temperature properties as compared to the corresponding monolithic materials used as the matrix. The exact nature of the GDS technique used in the fabrication of MMCs has a significant effect on the overall properties of the product. As shown earlier in this text, severe deformation of the particle interface due to shear instabilities must be achieved in order to obtain effective bonding between the matrix and the reinforcements.

Aluminum, copper, and nickel are the most common matrix elements employed in metal–ceramic composites; on the other hand, the most effective reinforcement particles are Al_2O_3, SiC, TiC, and others. For example, Conteras et al. (2004) reported that the addition of TiC as a reinforcement improves the mechanical properties at both moderate and high temperatures. The composites chosen for this study are shown in Table 8.2. It is widely recognized that the strength of the particle/matrix interface is of great importance with respect to the toughening of materials. In fact, the stiffness of the metal matrix is greatly increased by the presence of rigid particles. This has numerous additional advantages including an increase in both exploitation temperatures and thermal conductivity, as well as generating a larger coefficient of thermal expansion.

An important recent development in this area is concerned with the preparation of hybrid–particulate composites within which there are both soft and rigid particles. It has been found that under certain circumstances synergism can be obtained, with several toughening mechanisms operating together to

Tab. 8.1 Coating materials used for GDS according to Amateau and Eden (2000).

Powder material	Gas	Substrate
Aluminum	N_2	Metal, ceramic
Ti 35Zr 19Ni	N_2	Metal
Zinc	N_2	Metal, ceramic
Al Alloys + SiC(15%) (No. 12–17)	N_2	Metal
Cadmium	N_2	Metal
Al, Zn+10-15%HA (HA-Hydroxyapatite)	N_2	Metal
Tin	N_2	Metal
Al + 50Cu (Mixture)	N_2	Metal
Silver	N_2	Metal
Cu + 50Ni (Mixture)	N_2	Metal
Copper	N_2	Metal
Co32Ni 20Cr 8Al	N_2	Metal
Nickel	N_2	Metal
316St.Steel	N_2	Metal
Titanium	N_2	Metal
Ancorsteel 1000	N_2	Metal
Cobalt	N_2	Metal
Rhenium	N_2	Metal
Iron	N_2	Metal
Molybdenium	H_2	Metal
Niobium	N_2	Metal
Ni 23Co 20Cr 8.5Al 4Ta 0.6Y (Amdry, 997)	H_2	Metal
Al12Si	N_2	Metal
Inconel 718	H_2	Metal
Al6061	N_2	Metal
Diamalloy 1005 (Similar to Zuconel 625)	H_2	Metal
CCC (Al 9Ce 5Cr 2.8Co)	N_2	Metal
Ni-based superalloy AF 115 (Ni 15Co 10.7Cr 5.9W	H_2	Metal
Pech 1 (Al 12Zn 3Mg 1Cu 0.25Mn 0.2Cr 0.2Zn)	N_2	Metal
3.9Ti 3.8Al 2.8Mo1.7Ni 0.75H)	H_2	Metal
NYCZ (Al 14.5Ni 8.2Y 2.7Co 1.9Zr)	N_2	Metal
Ni 20 graphite (composite) (Amdry 954)	H_2	Metal
FCW (Al 9Fe 7Ce 0.4W)	N_2	Metal
Ni 45Cr 4Ti (Ametek 55)	H_2	Metal
Al Bronze (Cu 10Al, Cu9.5Al !Fe)	N_2	Metal
Al–Ni–Bronze	H_2	Metal
Cu 36Ni 5Zn	N_2	Metal
Hot Work Tool Steel AlSiH13	H_2	Metal
Cu 38Ni	N_2	Metal
WC-21 Co	H_2	Metal
Nichrome (80/20)	N_2	Metal
Co-106-1	H_2	Metal
Ni 5Al	N_2	Metal
Co 32Ni 21Cr 8Al 0.5Y	H_2	Metal
Cr3C2-25Ni CR	N_2	Metal
Co 29Cr 6Al 1Y (Amdry 920)	N_2	Metal

Tab. 8.2 Composite microstructure characteristics.

Composite designation	Particle designation	Average particle size (μm)	Porosity (%)	Volume fraction of reinforcement (%)	Interparticle distance (μm)
Al–Al_2O_3-1	Al_2O_3	10	3.1	10	187.5
Al–Al_2O_3-2	Al_2O_3	10	3.5	15	100.
Al–Al_2O_3-3	Al_2O_3	10	4.3	30	44.6
Al–Al_2O_3-4	Al_2O_3	45	4.9	50	10
Ni–TiC-1	TiC	20	7.5	10	90
Ni–TiC-2	TiC	20	8.1	20	57.1
Ni–TiC-3	TiC	20	9.5	30	33.3

create composites having very impressive mechanical properties. The mechanical behavior of GDS-deposited metals and MMC–particulate composites are discussed in detail below. Most of this discussion is concerned with Al, Cu, and Ni metal, as well as the comparison of MMCs created using both low and high-pressure GDS.

8.2.3 Definition of Structure Parameters

The main factors that need to be established to fully characterize the structure of particulate-reinforced MMCs are the size and the volume fraction of the particles. The size parameters of a material's microstructure can exert a strong influence on its mechanical properties even when all other parameters remain unchanged. An example of this influence can be seen as metallic matrixes are strengthened by particles when the generation and movement of dislocations are controlled. Kouzeli and Mortensen (2002) show that the analysis of such interaction phenomena has led to the development of a number of dislocation models which can adequately describe the correlation between mechanical properties and microstructural parameters in dispersion or precipitation hardened alloys.

The case of strengthening in particle-reinforced metal–matrix composites has been extensively studied in the past. Many dislocation models have been designed to account for strengthening in this class of materials. However, most of these models tend to underestimate the strengthening increment in the metallic matrix due to the presence of In reinforcement particles. Kouzeli and Mortensen (2002) state that this underestimation is due to the neglect of another, very different contribution to composite strengthening, namely load-sharing. Unlike precipitation or dispersion-hardened alloys, the volume fraction of the reinforcing phase in GDS composites is relatively high. Also, interparticle bonding areas fail to create a continuous network, as compared with regular MMC produced by infiltrating. In this case, the matrix sheds load to the reinforcement during straining. This concept of "load-sharing" is central to composite continuum mechanics, which predicts strengthening in particle-reinforced composites based on knowledge of the bulk metal matrix properties, the volume fraction, aspect ratio, and spatial arrangement of the particles. Analyses of GDS MMC properties from this viewpoint have not yet been performed; neither have the volume fraction of the reinforcement particles nor the spatial arrangement of the particles been adequately examined.

The volume fraction, ϵ_j, of particles can be calculated from the weights of the ingredients making up the composites by means of the following equation:

$$\epsilon_j = m_s \rho_m / (m_s \rho_m + m_m \rho_s) \tag{8.1}$$

where m_s is the mass of reinforcement particles, m_m the mass of matrix particles, ρ_s is the density of the particles, and ρ_m is the density of the matrix. As well as knowing the average volume fraction of particles, it is also essential to determine if there are any local variations in the composition due to factors such as the settling of particles during spraying. This can be monitored by polishing random sections taken from different areas of the coating.

The spatial arrangement of the particles in MMC is characterized by the mean free interparticle distance, λ. According to Underwood (1985), measurements of the mean free interparticle distance may be conducted by superposing random lines on micrographs of the as-deposited composites. λ may be determined from the number of particle intercepts per unit length of the test line, N_L , and the volume fraction of particles, ϵ_j, from

$$\lambda = \frac{1 - \epsilon_j}{N_L} \tag{8.2}$$

As-deposited composite microstructures feature a near homogeneous distribution of particles in an aluminum matrix with low porosity (Fig. 8.1(a)). Ni-based composites have a higher porosity, diminishing the mechanical properties. The basic microstructural features of the Al- and Ni-based GDS composites are summarized in Table 8.2. In all these composites, the reinforcement interparticle distance varies in the range of 10–200 μm, and is comparable with the soft matrix particle size (about 45 μm). Thus, the interparticle boundaries seem to exert a similar influence on dislocation creation and motion in the matrix, as compared to the reinforcement particles.

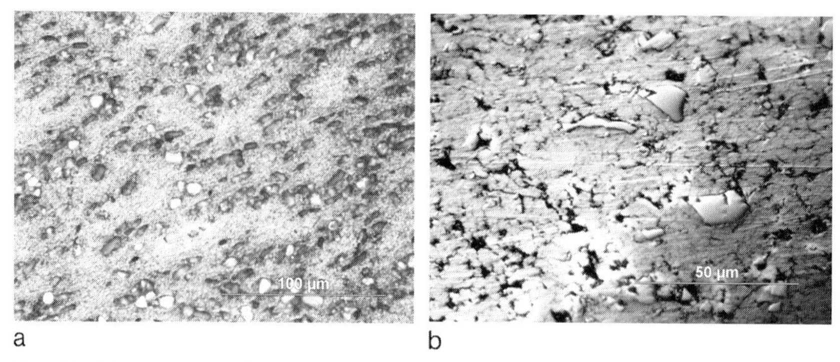

Fig. 8.1 Microstructure of Al–Al_2O_3 and Ni–TiC composites. (a) Microstructure of as-deposited GDS Al–15%Al_2O_3 composite; (b) microstructure of as-deposited GDS Ni–30%TiC composite.

8.3
Structure and Mechanical Properties of Composite Coatings

8.3.1
Methods of Testing

8.3.1.1 Strength Tests

Strength tests such as the tension test, the diametrical compression test, and the simple uniaxial compression test help determine the performance limits of a material with respect to stress. As such, strength criteria can be built using a combination of the local principal stresses. For brittle materials, these limiting stress states (Fig. 8.2) reflect a *combination* of the fracture toughness of the material and the presence of defects in the tested specimen. Such tests are applicable when we are interested in characterizing the fracture related mechanical performance of specimens in which exact knowledge of the defects and their interaction with the local stress is unknown (Procopio et al. 2003). The peculiarities of GDS composite microstructure, such as pores, interparticle bonding areas and type, and the spacing of the reinforcement particles define the composite mechanical performance. Thus, all of the various types of mechanical tests mentioned above are necessary to differentiate between these factors. In order to fully characterize the structural strength of composites it is necessary to employ three distinct types of compression tests to determine the fracture toughness of composite: uniaxial compression, diametrical compression, and diametrical compression of notched samples (Figs. 8.3(a) and (d)). Additionally, tension (ASTM Designation E8-01) and adhesion tests (ASTM 633) were employed to define bonding and adhesion strength.

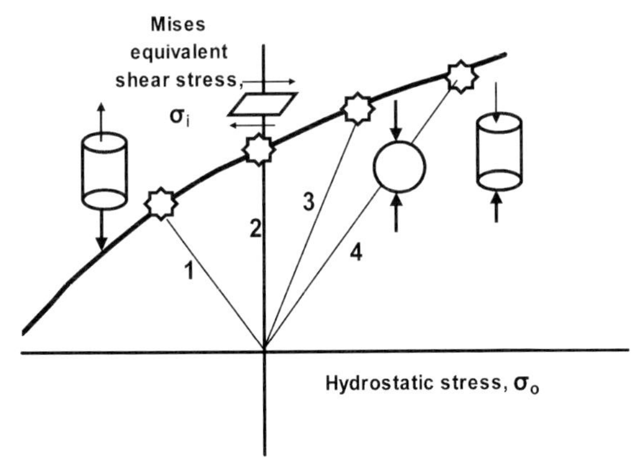

Fig. 8.2 Schematic representation of a failure locus near the shear stress dominant region for pressure-dependent yielding materials. $\sigma_0 = 1/3(\sigma_x + \sigma_y + \sigma_z)$ – hydrostatic (compressive) stress; $\sigma_i = (1/2)[(\sigma_x - \sigma_y)^2 + (\sigma_y - \sigma_z)^2 + (\sigma_x - \sigma_z)^2]^{1/2}$ – Mises equivalent shear stress, where σ_x, σ_y, and σ_z represent the principal directions of stress (after Procopio et al. 2003).

The cylindrical samples of $\oslash 15 \times 15$ mm were loaded in uniaxial compression using a mechanical testing machine (Zwick, Model T1-FR150TL). The displacement was recorded using an extensometer placed on the crosshead. The samples were loaded under compression at 0.5 mm/min until failure.

The diametrical compression test involved loading the cylindrical specimens across their diameter such that tensile stresses were created perpendicular to the load. Also known as the Brazilian disk test, this compressive test allows for the use of samples having simple disk geometries. Here the force limit that is necessary to bring about failure is measured and used to estimate the tensile strength σ_f

$$\sigma_f = \frac{2P_f}{\pi D t} \tag{8.3}$$

where P_f is the fracture load , D is the disk diameter ($D = 25$ mm), and t is the sample thickness ($t = 5$–10 mm). The simple geometry and loading conditions are attractive for materials which are too difficult to process or machine into the ASTM standard (ASTM Designation E8-01) specimen. It is usually assumed in this case that the tensile strength of the specimen can be expressed in terms of the maximum principal tensile stress in the sample (Procopio et al. 2003). The diametrical compression test is being used to estimate strength for elastically deforming materials, which are susceptible to brittle fracture.

The fracture toughness measurements that were made with notched diametrical compression were made with samples of identical size. The notch was aligned parallel to the loading axis. The test was performed using a fairly high crosshead speed of 0.5 mm/min in order to minimize time-dependent sub-critical crack growth. The fracture toughness was calculated as follows:

$$K_{1C} = 1.12\sigma_f\sqrt{\pi c} \tag{8.4}$$

$$K_{IC} = \frac{2.24}{\sqrt{\pi}}\frac{P_f\sqrt{c}}{Dt}$$

where c is the crack depth, D is the specimen diameter, and t is the specimen thickness.

For strength tests, ten specimens per condition were tested; for fracture toughness tests, four specimens per condition.

8.3.1.2 **Determining the Elastic Modulus**

The pulse-echo-overlap technique applied by Krautkramer and Krautkramer (1983) was used to determine the time of flight. This was then used to compute the ultrasonic velocity (V), which in turn was used to calculate the elastic constants using the relationship $C_{ij} = \rho V_{ij}^2$. The experimental setup for the ultrasonic system used in this study is discussed in detail by Maev (2007). Commercial ceramic transducers of 0.25-in diameter and 5 MHz fundamental frequency were used for both the longitudinal and transverse mode measurements. A sketch of the sample with respect to the spraying direction is shown in Fig. 8.4. The sample is considered to be transversely isotropic in the plane perpendicular to the spraying direction.

The remaining portion of the samples was used to determine the longitudinal velocity of sound. To allow for a proper reading the sample had to have parallel surfaces. As such, the samples were mounted to a metal sample using two sided tape and polished until both sides were smooth and parallel. The samples were then studied using an AM 1102 acoustical microscope (Tessonics) utilizing a 20 MHz flat transducer with a 3 mm diameter. Due to the porous nature of the material it was undesirable to immerse the samples and therefore Imagel R03-GEL1 (Tessonics) was used. By measuring the time t for a pulse to echo through the sample at a point of known thickness d the longitudinal velocity V is determined from

$$V = \frac{2d}{t}$$

8.3.1.3 **Preparation of Samples**

The composites were produced with a low-pressure GDS test rig (Fig. 5.1). The cylindrical samples of various sizes were built up using a special device which allowed uniform spray formation of the bulk material (Fig. 8.5). The device consisted of a bottom punch (which played the role of the substrate)

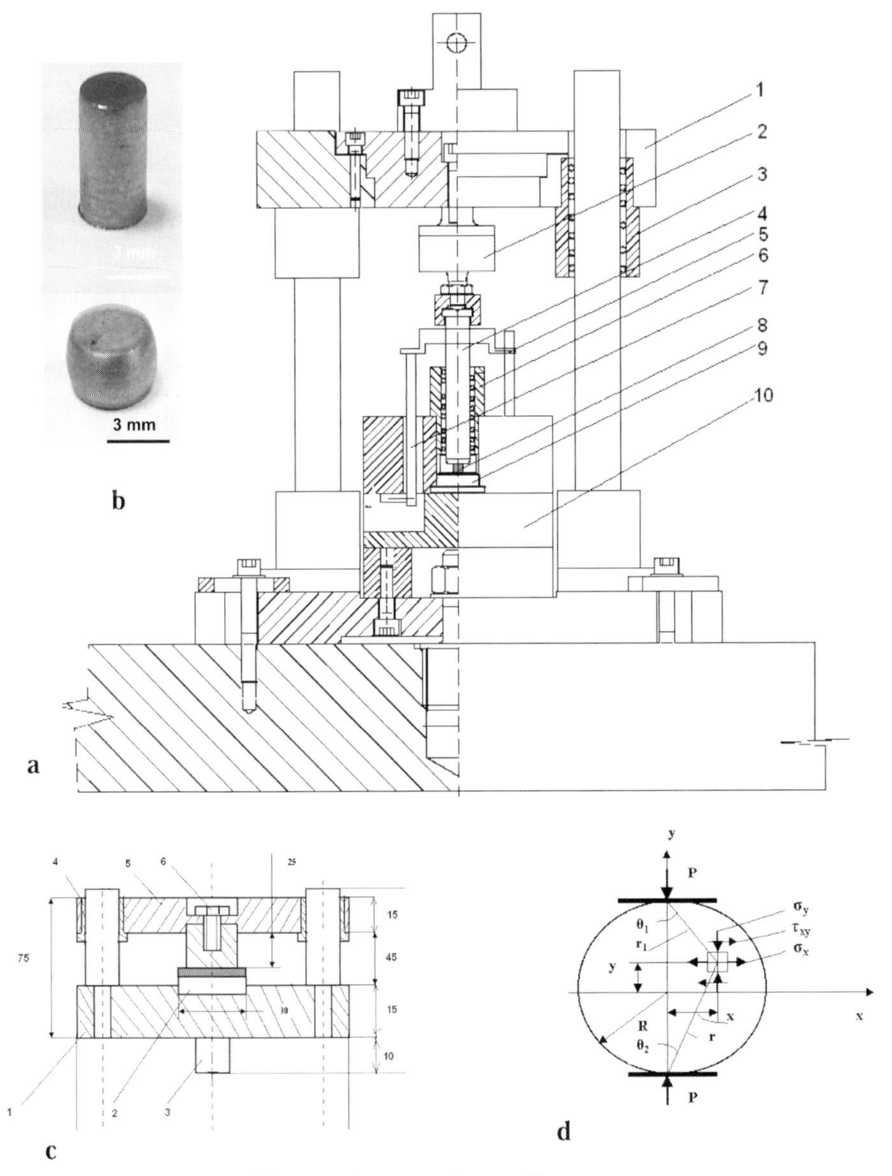

Fig. 8.3 Mechanical tests. (a) Uniaxial compression test fixture; (b) compression sample (before and after test); (c) adhesion test fixture; (d) diametrical compression test scheme.

and a die (which played the role of a mask). Cylindrical samples were obtained with high accuracy. The bottom punch was rotated with definite traverse speed and lowered with an axial velocity that equal to the deposition

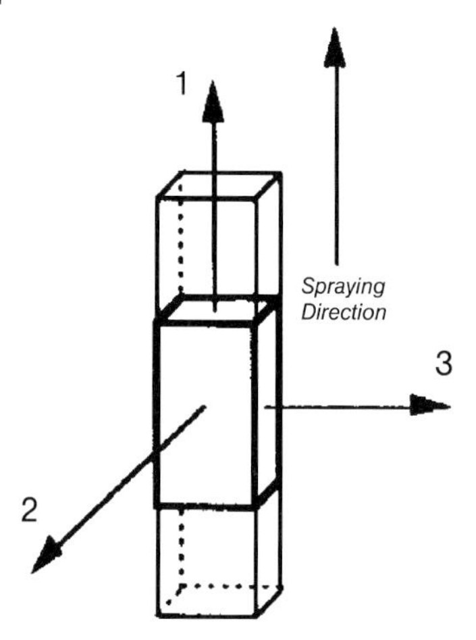

Fig. 8.4 A sample for elastic modulus measurement.

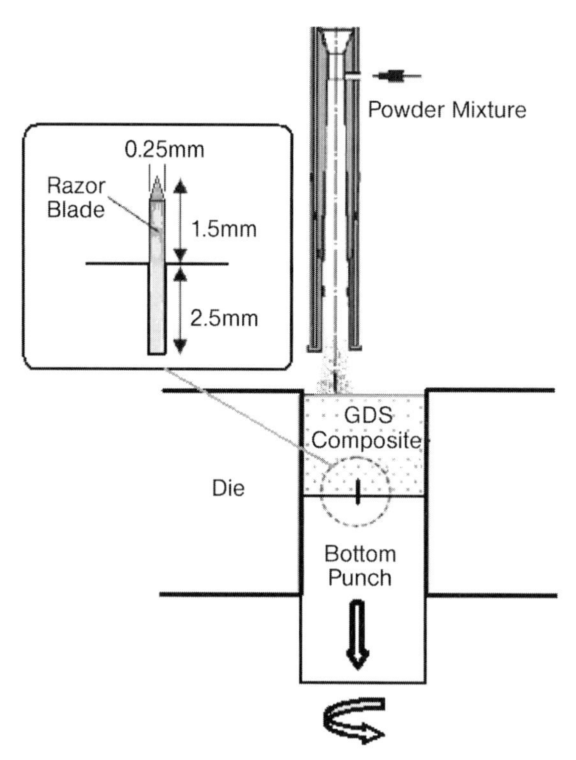

Fig. 8.5 Cylindrical sample preparation scheme.

rate. The bottom punch movement results in a constant stand-off distance and stable deposition parameters during the spray process.

First, Al and Ni and reinforcement powders Al_2O_3 and TiC were blended homogeneously. Next, the blended powders were sprayed with air temperature 400°C (Al-based composites) and 500–600°C (Ni-based composites). By varying the average size and chemistry of the reinforcing ceramic phase (Table 8.2), diverse composite microstructures were obtained. The angular Al_2O_3 and TiC powders of commercial grades were used to produce seven composites, which will be referred to hereafter as shown in Table 8.2.

8.3.2 Analysis of the Elastic Modulus

8.3.2.1 General Relationships

In order to increase the strength and stiffness of the composite materials, the reinforcement particle loading must be as high as possible. However, previous studies have shown that local reinforcement particle clustering might become a serious problem when the reinforcement loading is very high, e.g., at 30 vol.% (Tanga et al. 2004). Some ceramic reinforcement particles that may become clustered (e.g., SiC) usually cannot be sintered and bonded together at the consolidation temperature used for Al-based MMCs. Weak bonding between the ceramic particles may play a major role in microcrack initiation and growth, which can reduce the modulus of elasticity and fracture toughness of the MMCs. Therefore, as shown by Tanga et al. (2004), high-volume fractions of reinforcement particles (>30 vol.%) are not usually used in current structural composites. In the case of GDS composite coatings, the porosity has great potential influence on the mechanical properties and, as such, must be taken into account.

The addition of high-modulus particles to a low-modulus metal matrix invariably leads to an increase in the modulus E of the composite over that of the matrix. As shown in Fig. 8.6, there is a strong dependence of the modulus composites upon ϵ_j, for Al-based composite reinforced with Al_2O_3 particles. This behavior has been well documented by Orrhede et al. (1996), Peng et al. (2001), and Kouzeli and Mortensen (2002). Further, a variety of theories have been developed to explain the dependence of E upon ϵ_j, two of which are summarized below.

8.3.2.2 Rule of Mixture (ROM) Bounds

In this model, there are two bounds for the elastic constants of the composite. The upper bound for ROM assumes equal strain distribution in the composite. In this case, Young's modulus E is effectively given by

$$E = (1 - \epsilon_j)E_m + \epsilon_j E_r \quad (8.5)$$

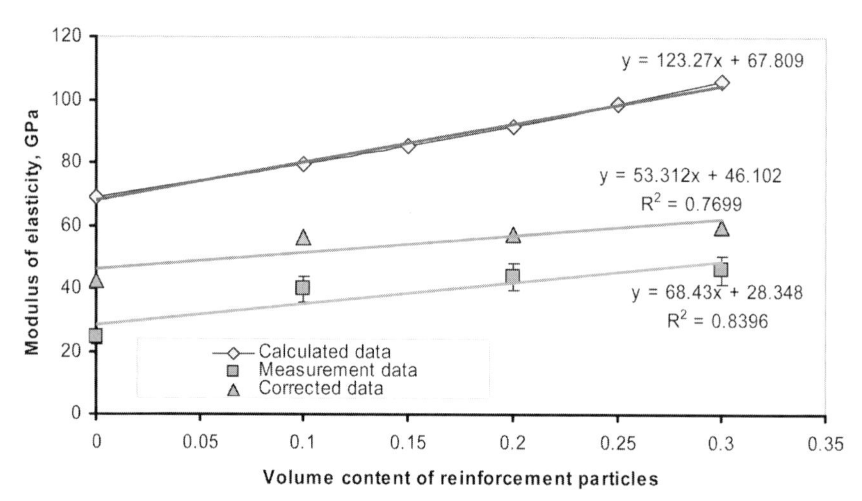

Fig. 8.6 Elastic modulus of Al-based composites.

where ϵ_j is the volume fraction of reinforcement, and the suffixes r and m represent reinforcement and matrix, respectively.

The lower limit of ROM is also obtained by assuming equal distribution in the composite. This is referred to as the ROM of compliance and is given by

$$\frac{1}{E} = \frac{(1-\epsilon_j)}{E_m} + \frac{\epsilon_j}{E_r} \tag{8.6}$$

In practice, Orrhede et al. (1996) have found that these bounds are too far apart to be of significant use. Nevertheless, their simple nature makes them useful for approximations.

8.3.2.3 **Hashin–Shtrikman (H–S) Model**

This model is based on the macroscopical isotropy and the quasihomogeneity of the composite, where the shape of the reinforcement is not a limiting factor (Green 1998). The model initially assumes a homogeneous and isotropic reference material wherein the constituents are dispersed. The fundamental Hashin–Shtrikman equation is

$$E_c = \frac{E_m[E_m(1-\epsilon_j) + E_r(\epsilon_j + 1)]}{E_r(1-\epsilon_j) + E_m(\epsilon_j + 1)} \tag{8.7}$$

8.3.2.4 **Effect of Porosity on Elastic Constants**

As stated by Green (1998), porosity leads to a decrease in the elastic modulus because the load-bearing area of a material is being reduced by the pores, with the stress becoming "concentrated" near the pores. For a body containing a low concentration of spherical pores, the MacKenzie (1950) solution for Young's modulus of the porous body is given by

$$E = E_0(1 - Af + Ff^2) \tag{8.8}$$

where E_0 is Young's modulus of the dense material, f is the porosity fraction, and A and B are constants of the order of 1.9 and 0.9, respectively.

The assumption that the elastic modulus of compacted powder is close to zero leads to the conclusion that the change in the elastic constants is simply given by the degree of densification:

$$\frac{E}{E_0} = \left(1 - \frac{f}{f_g}\right) \tag{8.9}$$

where f_g is the porosity fraction in the green compact. Thus, $E = 0$ when $f = f_g$ and $E = E_0$ when $f = 0$. However, Young's modulus is not zero even in the green body because adhesion between the powder particles and the use of binders give finite values for the elastic constants.

Young's modulus measurement results (Fig. 8.6) reveal the dependence $E = f(\epsilon_j)$ for LPGDS Al-based composites which is below as was calculated using the H–S model. An appropriate correction to the measurement data takes into account the composite porosity equation (8.8), resulting in an increase in E. It should be noted that the calculation of E with H–S equation (8.8) assumes that the structure of the GDS composite is similar to the infiltrated composites described by Kouzeli and Mortensen (2002).

For LPGDS, in accordance with the classification by Borchers et al. 2004, interparticle bonding may be characterized as adiabatic bonding. This adiabatic bonding leads to sound velocities similar to metallurgical bonding obtained in MMC casting or infiltrated MMC. However, as shown in Chapter 2, the area of adiabatic bonds in the region of interparticle contact constitutes only a fraction of the total area. This means that about 35–45% of the contact area is not adiabatically bonded, leading to an overall decrease of Young's modulus. In order to more precisely characterize the mechanism of interparticle bonding, then, it is necessary to analyze Young's modulus in the region of bonding.

In general, three approaches have been adopted to deal with mechanical property–porosity relations. These are: (i) mechanical strain analysis of bodies with generalized structure, (ii) an influence of pore shape and resulting stress concentration effects (SCE), and (iii) consideration of the actual geometrical load-bearing area. Recently, the last approach has been expanded by Rice (1996) suggesting that most physical properties, including mechanical properties (e.g., elastic modulus, strength, etc.), are determined by the minimum contact area (MCA) fraction normal to the stress. An estimation of MCA is of great importance to define the real interparticle bonding area. For the case of sintering, MCA will depend primarily on the initial geometry of particles in the compact and their packing pattern. For the case of GDS, MCA is believed to depend on the adiabatic shear band formation parameters for certain particles.

8.3.2.5 **Development of MCA Model for GDS Process**

The central feature of the GDS shock consolidation processes is that energy is dissipated by a shock wave as it traverses a powder medium. Interparticle adiabatic shear band formation, vortices, voids, and particle fracture may occur due to plastic deformation. Meyers et al. (1999) describe various energy dissipation processes which take place during shock consolidation: plastic deformation, interparticle friction, microkinetic energy, and defect generation. If we assume sintering to be an ideal bonding process and characterize bonding by the relation v/v_0, we may compare our ultrasonic velocity measurement data with theoretical calculations of Mizusaki et al. (1996) for sintering. In fact, Fig. 8.7 shows the sintering process according to a model based on the minimum contact area between particles by Mizusaki et al. (1996). In Fig. 8.7(a), the particles are initially touching without any deformation. The model uses two sine functions to approximate the particle shape at all times. The first sine function represents the interparticle distance (given as c). The second sine function represents the neck, or area of contact between subsequent particles. The neck is characterized by its diameter r_o and its thickness, $1 - c$ (see Fig. 8.7). Figure 8.7(b) shows that as sintering takes place, the interparticle distance will grow, as does the neck. The neck diameter will continue to widen until full density is reached, as in Fig. 3.7(c) where the diameter of the rod is given as $2a_t$. However, as Mizusaki et al. (1996) demonstrates, it is unnecessary to consider absolute values; in fact, a_t may be taken to equal $^1/_2$. Figure 8.7(c) does show that when full density is achieved, r_o is equal to $2a_t$. Figure 8.7(d) shows the result of this interpretation using relative distances. The first sine function represents the particle from 0 to c, while the second function goes from c to 1. The resultant functions devised by Mizusaki et al. (1996) are

$$f(x) = \frac{1+r_0}{2} + \left(\frac{1-r_0}{2}\right) \sin\left(\frac{\pi x}{c}\right) \quad \text{for } 0 \le x < c \tag{8.10}$$

and

$$f(x) = \frac{1+r_0}{2} + \left(\frac{1-r_0}{2}\right) \sin\left(\frac{\pi(x+1-2c)}{1-c}\right) \quad \text{for } c \le x < 1 \tag{8.11}$$

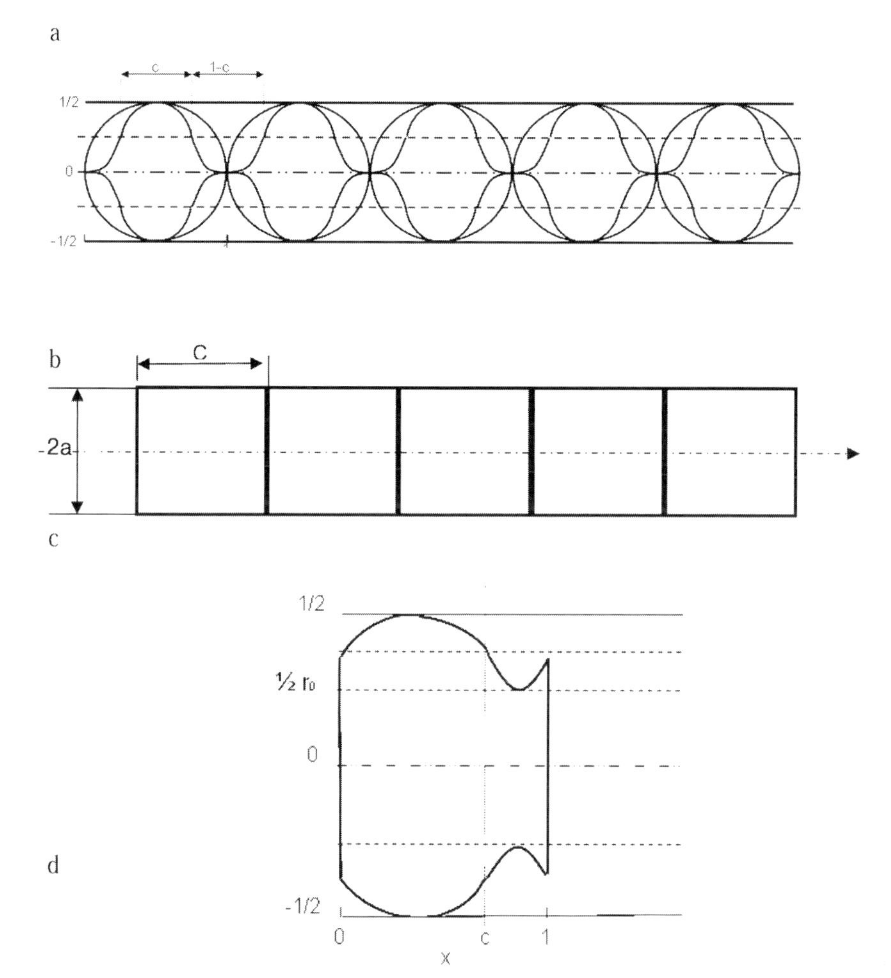

Fig. 8.7 String model of sintering, after Mizusaki et al. (2001). (a) A string of particle spheres and its approximation with two functions; (b) the sintered string of particle spheres with neck diameter r_o; (c) full density sintered particle string; (d) the view of one spherical particle after sintering with neck diameter r_o.

According to their interpretation, this gives a relative density of

$$\frac{\rho}{\rho_0} = \int_0^c \left[\frac{1+r_0}{2} + \left(\frac{1-r_0}{2}\right)\sin\left(\frac{\pi x}{c}\right)\right]^2 + \int_c^1 \left[\frac{1+r_0}{2} + \left(\frac{1-r_0}{2}\right)\sin\left(\frac{\pi(x+1-2c)}{1-c}\right)\right]^2 dx \tag{8.12}$$

when compared with full density, ρ_0. It should be noted that this model may be obtained using only the relative length rather than absolute values. A "particle column" model of GDS shock consolidation is shown in Fig. 8.8. In this

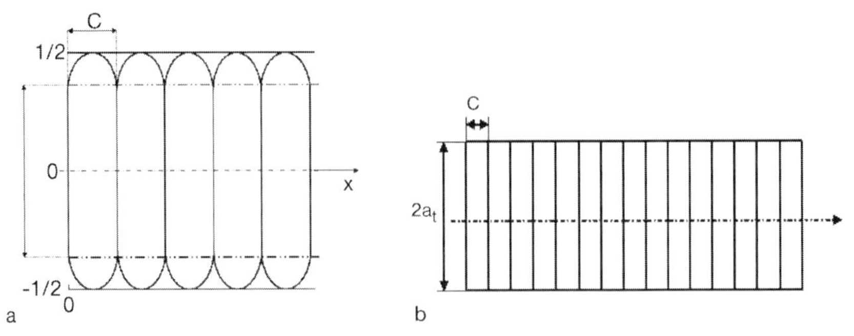

Fig. 8.8 GDS model of densification. (a) The string of spherical particles after impact consolidation during GDS; (b) full density consolidated particle string due to GDS.

case only particle deformation is responsible for the increase in neck diameter. The thickness of adiabatic shear bands for the GDS case is considered to be several microns (Maeva et al. 2006). For this reason, the thickness of the neck is approximated to be zero, and, consequently only the first integral is required to model the particles. However, it is necessary to modify this function to fit the boundary conditions: When $x = 0$ or $x = c$,

$$F(x) = r_0/2 \tag{8.13}$$

By adapting Eq. (8.11) to meet these conditions one obtains

$$F(x) = \frac{r_0}{2} + \left(a_t - \frac{r_0}{2}\right) \sin\left(\frac{\pi x}{x}\right) \tag{8.14}$$

Finally, by putting this function into the relative density, we yield

$$\frac{\rho}{\rho_0} = \int_0^c \left(r_0 + \left(a_t - \frac{r_o}{2}\right) \sin\left(\frac{\pi x}{c}\right)\right)^2 dx \tag{8.15}$$

In this case the model can be obtained using absolute values rather than the relative length.

Another notable feature of the GDS shock consolidation model is that the particle deformation process is one in which the volume of the particle is unchanged. Thus, at any moment of deformation the solid material incompressibility equation $V_o = V_i$ is valid. When $a_t = a_o$, the relative density of the particle string (Fig. 8.7(a)) is the ratio V_o/V_{io}, where $V_{io} = \pi a_0^2 2a_o$.

The pore fraction may be calculated based on a particle string representative element (Fig. 8.7(a)):

$$\theta = \frac{V_{\text{pore}}}{V_{\text{cyl}}} = \frac{1 - 4a_o}{3a_t^2 c} \tag{8.16}$$

where a_t is a radius of circumscribed cylinder. Taking into account $\frac{\rho}{\rho_0} = 1 - \theta$ the relative density is

$$\frac{\rho}{\rho_0} = \frac{(ma_t + nr_o)^2}{16a_t^2} \tag{8.17}$$

where m and n are the coefficients of approximations which fit the following boundary conditions:

1. if $a_t = a_o, r_o = 0$ and $\frac{\rho}{\rho_0} = 0.33 - 0.6$, and
2. if $2a_t = r_o, \frac{\rho}{\rho_o} - 1$

Further adaptations have been made to this theory using results from the works of Mukhopadhyay and Phani (1998, 2000). They approximated the normalized minimum contact area to be equal to the relative elastic modulus of the sample. The minimum contact area (MCA) is given by the smallest neck diameter. For the case of interest, this occurs when $F(x) = r_0/2$, giving MCA $= \pi r_0^2/4$. The normalized minimum contact area (NMCA) is thus r_0^2/a_t^2. For all practical purposes, as shown by the Martin and Rosen (1997), the NMCA is effectively equal to the relative modulus of elasticity E/E_0:

$$\text{NMCA} = \frac{r_0^2}{a_t^2} \equiv \frac{E}{E_0} \tag{8.18}$$

The GDS process results in real adiabatic bonds that are achieved within smaller contact areas than NMCA. Thus, the difference between theoretical and experimental obtained values for the ratio E/E_0 appears to result from interparticle adiabatic bonding at GDS.

8.3.2.6 **Elastic Modulus and Microstructure of LPGDS Composites**

The microstructures of Al and Al–10%Al_2O_3 GDS coatings (shown in Figs. 8.9(a) and (b)) reveal that the porosity of the Al coating is significantly larger than that of Al–10%Al_2O_3. Also, an indication of severe plastic deformation of particles may be clearly seen. It is possible to approximate the degree of interparticle contact through their evaluation using an optical microscope. The interparticle boundaries with adiabatic bonds are believed to be more stable for etching than those with only adhesion (Figs. 8.9(a) and (b)). The thin grain boundaries within the Al particles may be seen in Fig. 8.9(a). It seems to be reasonable that interparticle boundaries of the same thickness (for example those shown by arrows (A)) reveal adiabatic bonding, while thicker ones reveal only adhesion bonding (shown by arrows (B)). Unfortunately, a quantitative evaluation of the real adiabatic joining area is not possible through metallographic procedures. Neither is the analysis of the fracture topography

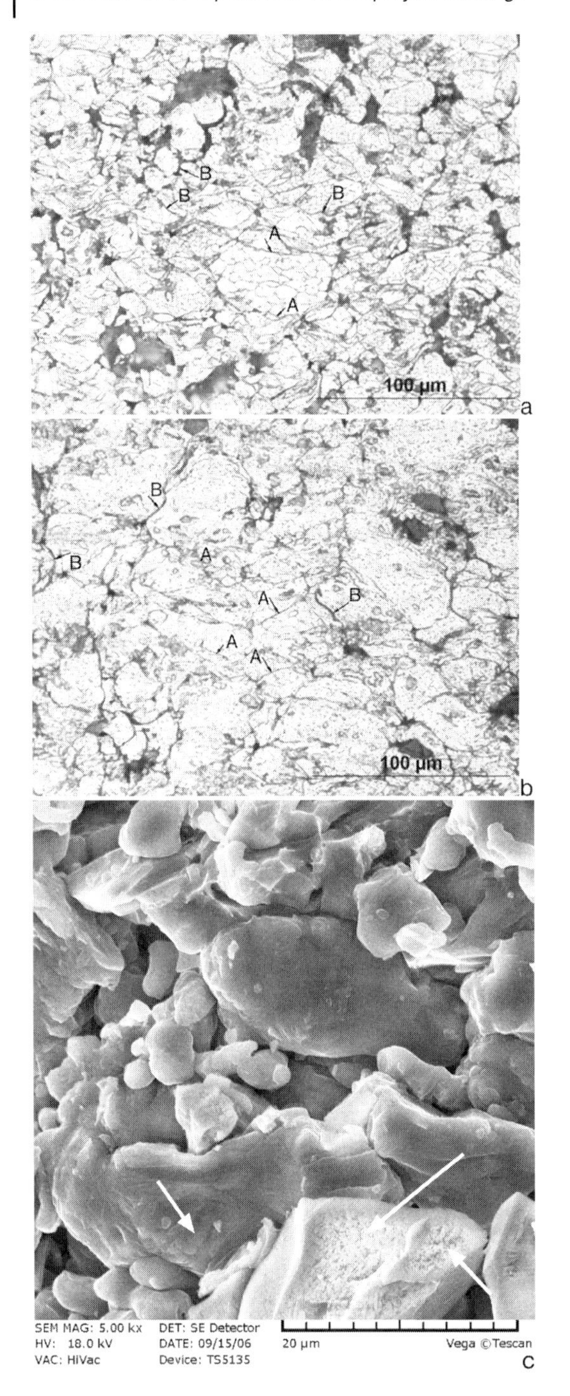

Fig. 8.9 Microstructure of the low-pressure GDS coatings. (a) Al; (b) Al +10%Al_2O_3 (both are etched with 10% HCl in alcohol); (c) SEM of Al+10%Al_2O_3 fracture surface.

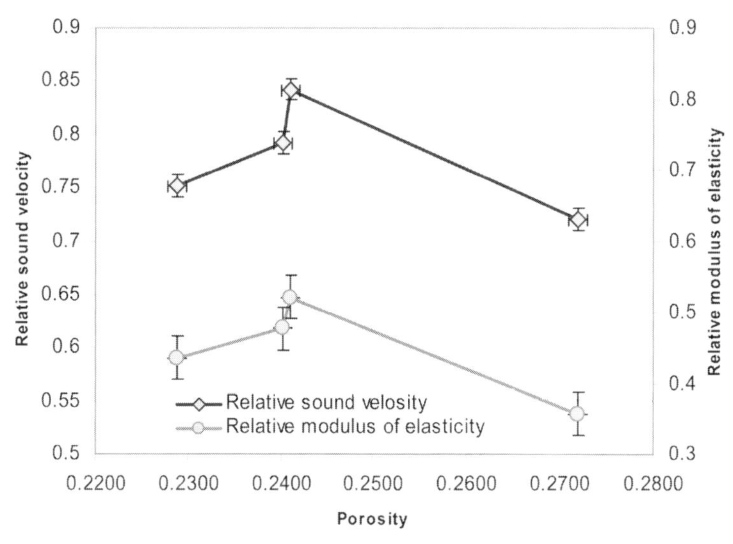

Fig. 8.10 Sound velocity of LPGDS sprayed Al composites.

useful for evaluation bonding, even in spite of the fact that areas of adiabatic bonding are clearly seen on the SEM images (Fig. 8.9(c), shown by arrows).

A quantitative analysis of the adiabatic bonding may be made by comparing the experimental and theoretical data of NMCA based on models (8.16) and (8.18). The results of the sound velocity measurement are shown in Figs. 8.10 and 8.11. The two main factors that influence sound velocity of the material are density and composition. Also, the densification process during spraying powder depends on the composition of the powder mixture. An increase in the Al_2O_3 content results in a decrease in the porosity from 0.272 to 0.229; this is due to an intensified plastic deformation of the soft Al particles. This effect is significant only for relatively small contents of Al_2O_3 (up to 10–15%). Further, increases in the content of Al_2O_3 beyond 25–30% result in an increase in porosity due to the considerable change that takes place with respect to interparticle friction. Al_2O_3 in the shock compaction of a powder layer can significantly affect sound velocity and elasticity modulus parameters (Fig. 8.10). This is due to (i) a decrease in porosity, and (ii) severe local deformation of the soft Al particles in the vicinity of interparticle contact (Fig. 8.9(c), shown by arrows).

The obtained data reveal a change in the mechanism of bond formation at porosities of about 0.2–0.25 (Fig. 8.12). The values of real NMCA are quite near to those obtained by sintering at high porosities. This means that severe deformation at interparticle contact results in real adiabatic bonding of particles along the entire contact surface. It is for this reason that the NMCA parameter (Eq. (8.18)) is measured to be quite near theoretical values calculated in accor-

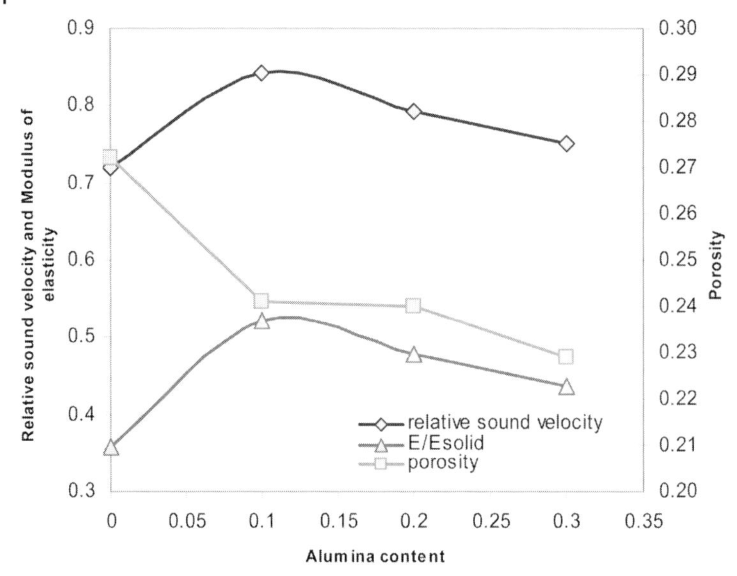

Fig. 8.11 Relative sound velocity, modulus of elasticity, and porosity dependences on alumina content in Al/Al_2O_3 coating composites.

dance with the Mizusaki model (Fig. 8.12). The GDS shock consolidation of higher density powder layers results in an increase in the contact area. However, adiabatic bonding is not achieved at all points. For this reason the values of NMCA do not vary considerably (Fig. 8.12, approximation of experimental data). A simple explanation of this effect is based on the adiabatic shear band (ASB) formation process during GDS shock wave consolidation. In the case of porous powder layer formation, the interparticle contact area is much smaller than for one of high density. The contact stresses, strains, and strain rates are known to be much higher in this case and as such ASB formation seems to be a highly probable event. When the contact area increases the stresses and strains should decrease. For smaller values of stress and strain, ASB formation is not enough, and the formation of the adiabatic bonds is not achieved. In fact, for porosities in the range of 0.03–0.23, the NMCA is about 0.4–0.42 for $Al–Al_2O_3$ coatings (Fig. 8.12). The main characteristic of GDS shock consolidation for small densities ($\theta = 0.25–0.5$) is the localization of deformation, resulting in high values of NMCA.

A validation of this model is shown in Fig. 8.13. One can see that the suggested shock consolidation model corresponds to experimental data in the range of $\theta = 0.25–0.5$ when the coefficients m and n of approximation (8.17) fit initial porosity $\theta_0 = 0.6$. A calculation of NMCA has been made using both the shock consolidation model described above and the Mizusaki model. This calculation yields similar results for $\theta_0 = 0.6$, however the curves for $\theta_0 = 0.25–0.5$ differ considerably. One can assume that this behavior is the

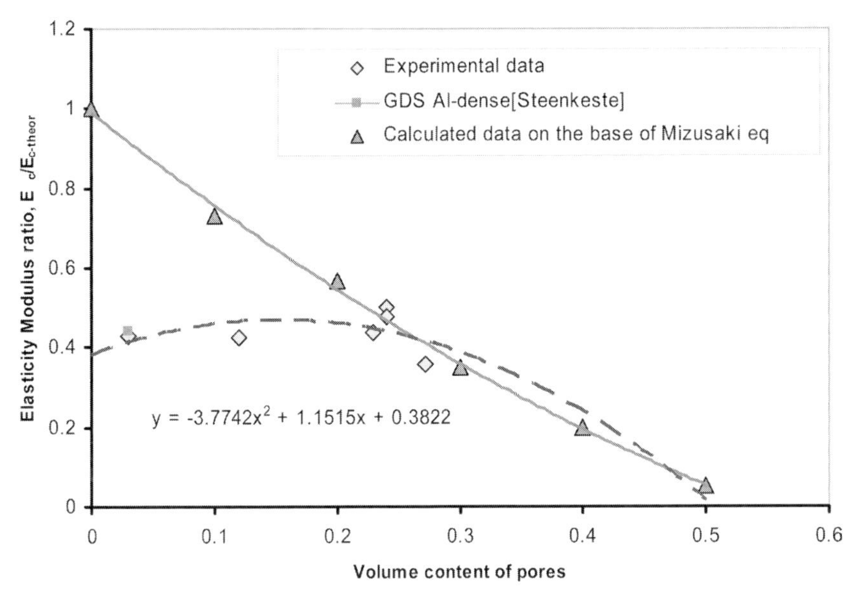

Fig. 8.12 Comparison of normalized elastic modulus of Al-based composite with Mizusaki equation.

result of localization effects at higher porosities – indeed, in the last case of higher initial porosity the number of interparticle contacts is reduced. That leads to an increase in the contact loads and strain rates at the interface between particles.

The numerical modeling results for the impact of Al particles ($d_p = 60$ μm) (Fig. 2.14) reveal that the radial velocities are higher than critical velocities only in a small region near the interface. Thus, in spite of the fact that tight contact is achieved, real adiabatic bonding may only be attained in this area. For this reason the NMCA value does not completely characterize the area of adiabatic bonding.

A similar behavior in the NMCA function may be observed for Ni–TiC composites (Figs. 8.14 and 8.15). Experimental results show that the real modulus of elasticity for LPGDS Al–Al_2O_3 and Ni–TiC composite coatings is only about 40–67% that of the solid material. This suggests that powder layer shock densification due to LPGDS gives rise to MCAs similar to that of HPGDS (Van Steenkiste et al. 2002).

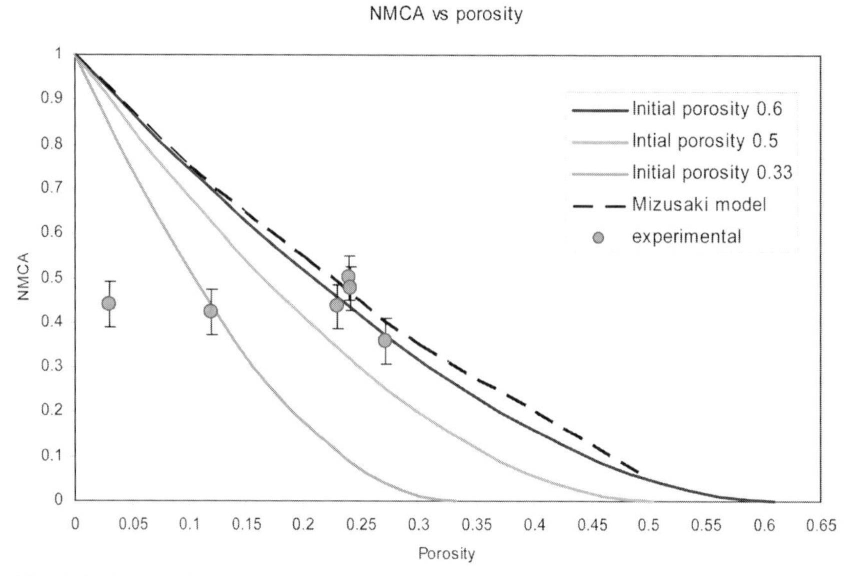

Fig. 8.13 Comparison of normalized minimum contact area (NMCA) of Al-based composite with GDS consolidation model.

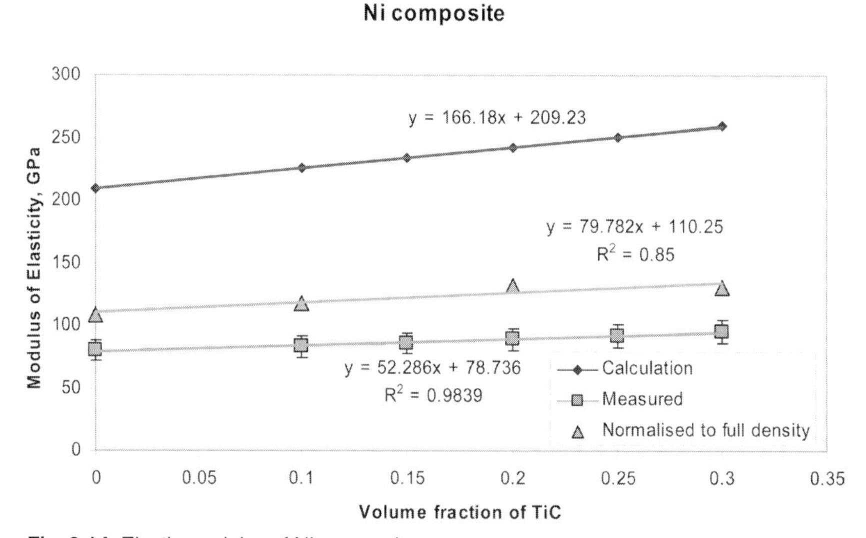

Fig. 8.14 Elastic modulus of Ni composite.

8.3.3
Load-Deformation Behavior of GDS Composites

8.3.3.1 Strengthening GDS composites

The deformation behavior of GDS aluminum coatings was studied by Van Steenkiste et al. (2002). It was found that even such inherently brittle mate-

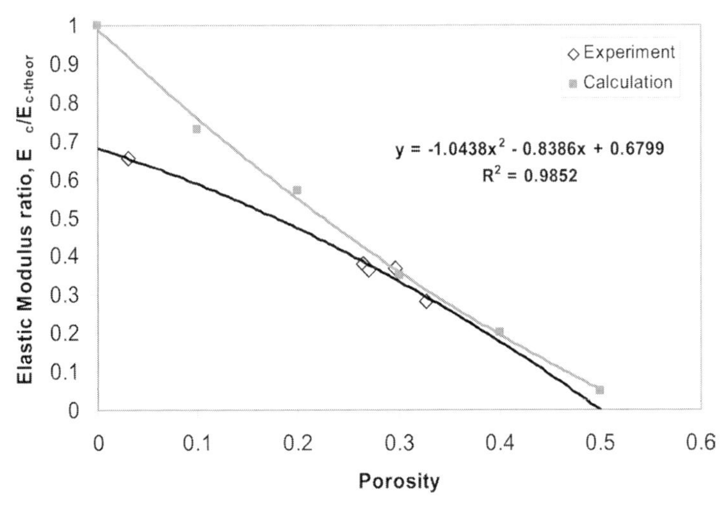

Fig. 8.15 Comparison of normalized elastic modulus of Ni-based composite with Mizusaki equation.

rials as GDS coatings are capable of undergoing a certain amount of plastic deformation by means of tension. A comparison of the mechanical properties of HPGDS Al coatings has been completed by Van Steenkiste et al. (2002), with the resultant data appearing in Table 8.3. The data reveal that, except for the modulus of elasticity (as discussed above), the mechanical properties of Al coatings are slightly lower than those of the bulk material. The effect of intensive deformation of particles during impingement with the substrate is perhaps obvious. However, it is interesting that the lack of interparticle bonding results in lower values of both yield point and ultimate strength. One must note that the application of low-pressure GDS process leads to an increase in coating porosity. This is why the LPGDS yield point is lower than that of HPGDS.

Tab. 8.3 Elastic modulus and tensile strength of Al.

	Material porosity (%)	Elastic modulus (GPa)	Yield point (MPa)	Ultimate tensile strength (MPa)
Bulk Al (Rothbart 1996)		69–72	35	90
Annealed Al (American Physics Handbook, 1972)		69	12	47
Cold rolled Al (American Physics Handbook, 1972)		69	106	112
HPGDS coating (Van Steenkiste et al. 2002)	3.0	30.8	56	75
LPGDS coating (authors results)	12	25.2	48	63

As shown previously, the application of powder mixtures for LPGDS coatings is believed to be one of the best solutions for the optimization of the GDS process, helping to achieve effective MMC coating structure and properties. However, neither the mechanical properties of LPGDS MMC coatings, nor the mechanisms of deformation have yet been studied. For this reason, the authors examined the deformation behavior of Al-based MMC coatings through the uniaxial compression of cylindrical samples of $\oslash 3 \times 6$ (see Fig. 8.3). This work yielded strain measurements in the range of $\epsilon = 0.001$–0.05.

These experimental results, coupled with data from related literature data are shown in the Figs. 8.16–8.18. These plots depict the characteristic features of flow stress σ_z against axial strain ϵ_z of Al and Al-based composites for small strains ($0.001 < \epsilon_z < -0.05$). It seems to be reasonable to compare the deformation behavior of Al–Al_2O_3 composites that were created using alternative technologies such as powder metallurgy (Narayanasamy et al. 2005) and infiltration (Kouzeli and Mortensen 2002). Figure 8.16 shows the flow curves of tension or compression of pure Al and Al–Al_2O_3 composites after sintering or spraying. The tensioning of HPGDS Al coating results in relatively small elongations (0.6%) because of the weak interparticle bonding of Al particles. The flow curves of pure Al (curves 1 and 5, Fig. 8.16) reveal similar values for flow stress, evidence that there are similar deformation mechanisms for both sintered and sprayed Al. Nevertheless, the hardening coefficient of sprayed material is higher than for sintered Al. The tension flow curves of infiltrated Al–Al_2O_3 composites (Kouzeli and Mortensen 2002) exhibit two stages of strengthening (curves 3 and 4, Fig. 8.16). The comparison of these flow curves with those of LPGDS materials (Figs. 8.17 and 8.18) reveals the following features:

1. The flow curve of LPGDS Al–10%Al_2O_3 composites lies near the tension curve of HPGDS Al;
2. The flow curves of LPGDS Al–10%Al_2O_3 and Al–15%Al_2O_3 composites are quite similar, and the influence of Al_2O_3 particle reinforcement is clearly seen;
3. The mechanisms of strengthening in the LPGDS Al–15%Al_2O_3 composite materials seem to be the same as those of the infiltrated composites, while the increase of Al_2O_3 content results in a change of deformation behavior in the microstrain region (up to $\epsilon = 0.005$);
4. The technology of making composites does not greatly influence the flow stress for strains higher than $\epsilon = 0.005$. The lower values of the flow stresses in this region are believed to be a result of the higher porosity and lower content of the reinforcement phase in the LPGDS composite materials, as compared with infiltrated ones.

The flow stress of the particle-reinforced composites, depends strongly on the mean interparticle distance λ, as defined by Eq. (8.2). Kouzeli and Mortensen

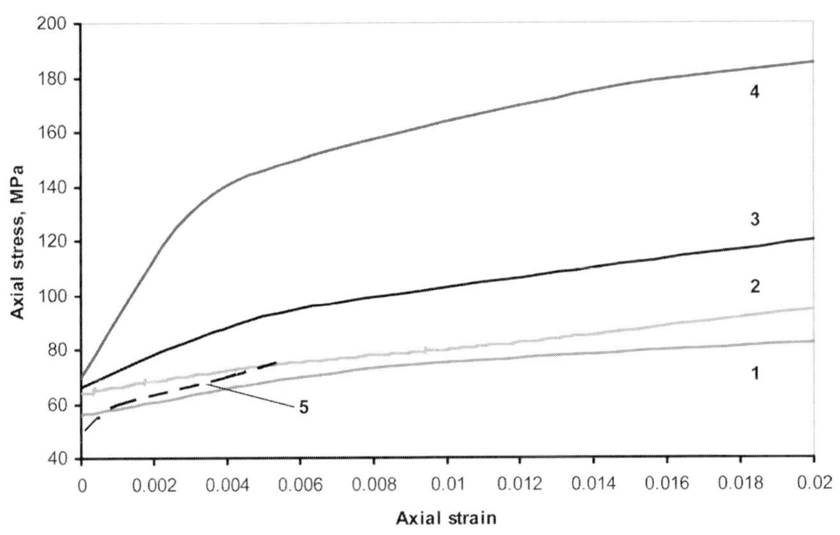

Fig. 8.16 Flow curves of Al-based compacts. (1) Compression of sintered Al (compacted and sintered 500°C, 1 h, $\rho = 97\%$, after Narayanasamy et al. (2005)); (2) compression of sintered Al–15%Al_2O_3 composite (compacted and sintered 500°C, 1 h, $\rho = 95\%$, authors results); (3) tension of infiltrated Al–46% Al_2O_3 composite, after Kouzeli and Mortensen (2002); (4) tension of infiltrated Al–54%Al_2O_3 composite, after Kouzeli and Mortensen (2002); (5) tension of cold sprayed Al (HPGDS, air ,$T = 288$°C, $\rho = 97\%$, after Steenkviste et al. (2002).

(2002) use the concept of geometrically necessary dislocations to evaluate this effect of size on the yield behavior of the composites.

During the plastic deformation of a crystal, dislocations are generated, move, and are stored. The accumulation of dislocations results in a dislocation density ρ_s, bringing about the phenomenon of work hardening. The dislocations that are stored due to incompatibility in deformation are called geometrically necessary dislocations, and the density of these dislocations is represented by ρ_G. As shown by Kouzeli and Mortensen (2002), the minimum density of geometrically necessary dislocations, p_G, in the deforming material can be linked to the average strain gradient η (in shear strain γ_m) present in its microstructure. Through deformation compatibility arguments they obtain

$$\rho_G = \frac{1}{b}\eta = 1\frac{\gamma^m}{b\hat{\lambda}} \tag{8.19}$$

where b is the magnitude of Burger's vector, and $\hat{\lambda}$ is the local length scale of the deformation field.

In the case of particle-reinforced metals (Tang et al. 2003), the incompatibility in deformation between the plastically deforming matrix and the rigid particles results in strong strain gradients in the metallic matrix. Thus, a finer

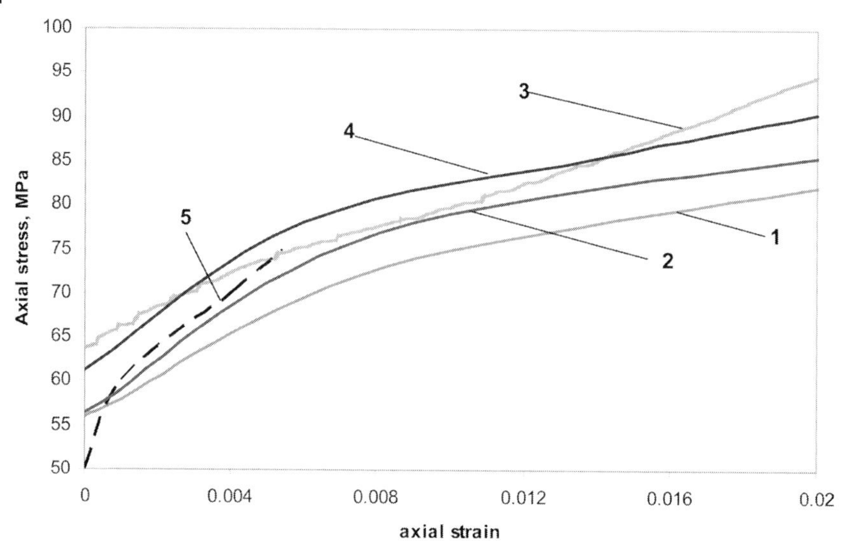

Fig. 8.17 Comparison of the flow curves of sintered and sprayed Al–Al_2O_3 compacts. (1) Compression of sintered Al (compacted and sintered 500°C, 1 h, $\rho = 97\%$, after Narayanasamy et al. (2005)); (2) compression of LPGDS Al–10%Al_2O_3 composite (LPGDS air 300°C, $\rho = 92\%$, authors results); (3) compression of sintered Al–15%Al_2O_3 composite (compacted and sintered 500°C,1 h, $\rho = 95\%$, authors results); (4) compression of LPGDS Al–15%Al_2O_3 composite (LPGDS air 300 °C, $\rho = 94\%$, authors results); (5) tension of cold sprayed Al (HPGDS, air , $T = 288$°C, $\rho = 97\%$, after Van Steenkviste et al. (2002).

composite microstructure should lead to a greater strain gradient in the composite matrix, which, in turn, should result in a greater density of geometrically necessary dislocations and higher composite flow stresses. Based on the previous analysis, the equation

$$\left(\frac{\sigma^m}{\alpha G^m b}\right)^2 = \rho_S + \rho_G = \rho_S + \left(\frac{4\epsilon^m}{b}\right)\frac{1}{\lambda} \tag{8.20}$$

represents the dependence of the matrix flow stress parameter $\left(\frac{\sigma^m}{\alpha G^m b}\right)^2$ on the dislocation density and microstructure characteristics of particle-reinforced composites. Here G_m is the matrix shear modulus, λ is the mean free interparticle distance, and ϵ^m is the average strain of matrix for the flow stress σ^m.

The results of assessment of the matrix flow stress parameter from the experimentally measured composite flow stress are shown in Fig. 8.19. Here the constants determined by Kouzeli and Mortensen (2002) were used. It can be seen that the effect of mean free interparticle distance on the matrix flow stress parameter is higher for LPGDS composites than for infiltrated composites, in spite of their lower density. This unusual behavior of LPGDS composites may

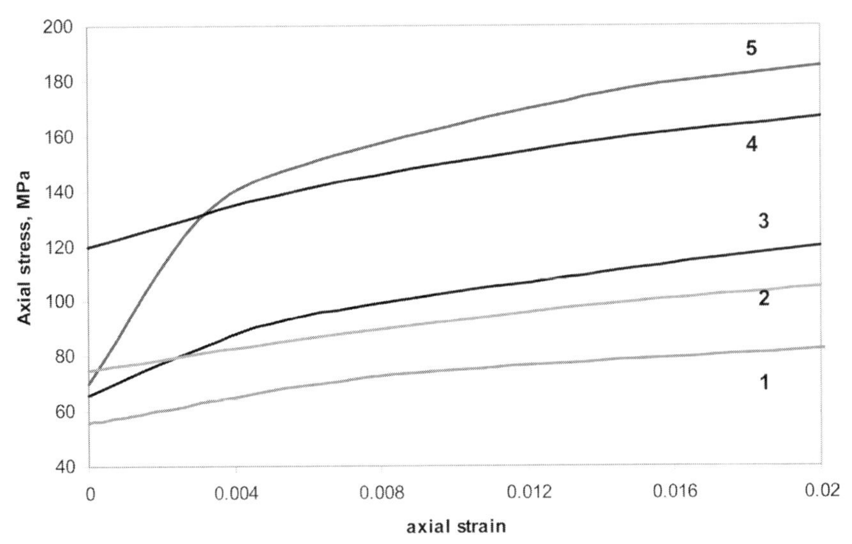

Fig. 8.18 Comparison of the flow curves of sintered and sprayed Al–Al_2O_3 compacts. (1) Compression of sintered Al (compacted and sintered 500°C, 1 h, $\rho = 97\%$, after Narayanasamy et al (2005)); (2) compression of LPGDS Al–30%Al_2O_3 composite (LPGDS air 300°C, $\rho = 93\%$, authors results); (3) tension of infiltrated Al–46% Al_2O_3 composite, after Kouzeli and Mortensen (2002); (4) compression of LPGDS Al–50%Al_2O_3 composite (LPGDS air 300°C, $\rho = 95\%$, authors results); (5) tension of infiltrated Al–54%Al_2O_3 composite, after Kouzeli and Mortensen (2002).

be attributed to the higher density of stored dislocations in these composites. These dislocations are due to the presence of adiabatic shear band areas which act as dislocation barriers. Also, the small grain size of the matrix is due to the severe deformation particles undergo during high velocity impact associated with the spraying process. For the same reason the yield stress 0.001 of LPGDS composites is higher then those of infiltrated composites.

The results of calculation (presented in Fig. 8.19) reveal the linear dependence of the matrix flow stress parameter on the reciprocal of the interparticle distance. This data is compatible with Eq. (8.20), which is based on peculiarities in the storage of dislocations in particulate-reinforced composites, which may in turn be compared with the classical Hall–Petch equation

$$\tau = \tau_0 + k_y d^{-\frac{1}{2}} \tag{8.21}$$

This comparison is justified in the case where there is an accumulation of dislocations against grain boundaries, resulting in dependence of the hardening increment on the square root of the grain size, d (Arzt 1998):

$$\tau = \frac{k_y}{\sqrt{d}} \tag{8.22}$$

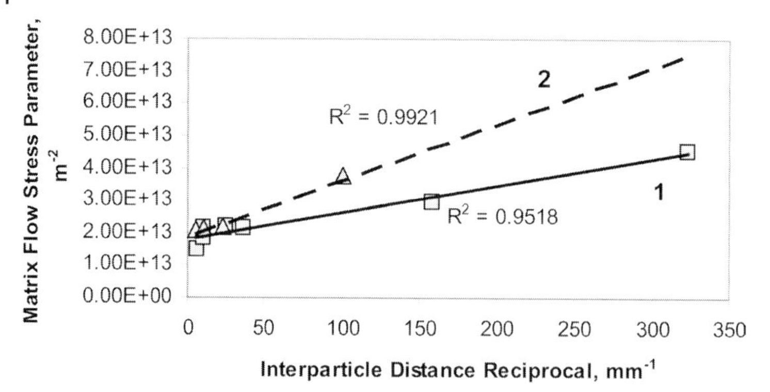

Fig. 8.19 Matrix flow stress parameter approximations for the total flow stress $\sigma_{0.01}$. (1) Tension of infiltrated Al–Al_2O_3 composites, after Kouzeli and Mortensen (2002); (2) compression of LPGDS Al–Al_2O_3 composites (LPGDS air 300°C, $\rho = 95\%$, authors results).

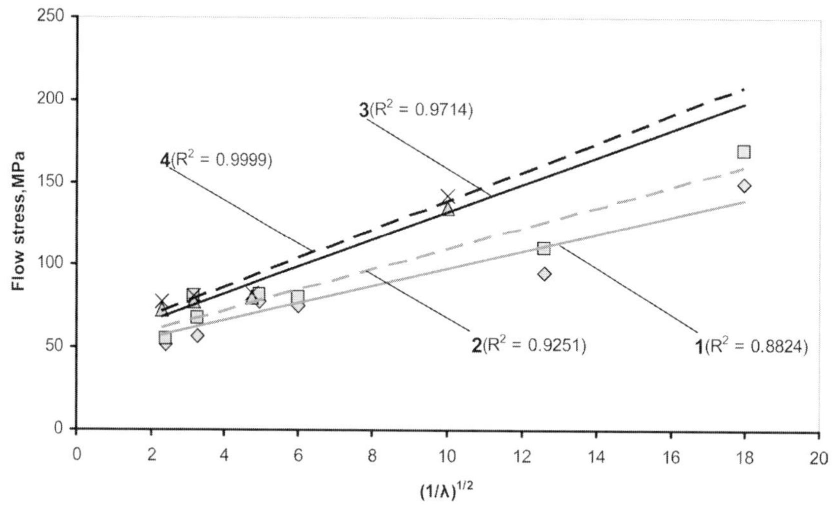

Fig. 8.20 Hall–Petch approximations (1), (2) $\sigma_{0.005}$ and $\sigma_{0.01}$ of infiltrated Al–Al_2O_3 composites, after Kouzeli and Mortensen (2002); (3), (4) $\sigma_{0.005}$ and $\sigma_{0.01}$ of LPGDS Al–Al_2O_3 composites (LPGDS air 300°C, $\rho = 95\%$, authors results).

Here τ is the flow shear stress and τ_0 and k_y are constants. Also, an assumption is made that the pile-up of dislocations is influenced by the mean free interparticle distance λ. Hall–Petch approximations of experimental results are shown in Fig. 8.20. The data reveal that the simulation veracity of linear approximations is better for sprayed composites than for infiltrated ones. Thus, the Hall–Petch pile-up model for strengthening LPGDS composites seems to be quite reasonable.

8.3.4
Failure Criterion and Microstructural Aspects of Crack Propagation

It is well known that GDS deposited materials tend to be brittle in their as-sprayed state. This is because the particles are work hardened during the deposition process and interparticle bonding is incomplete. The classical energy approach developed by Griffith (1920) seems to be useful for analyzing the bonding properties of GDS composites.

On the basis of linear elastic fracture mechanics (Anderson 1995), the elastic energy per unit crack area, G, that is available for an infinitesimal crack extension may be determined from the change in compliance of the cracked body as the crack extends by a small amount (Green 1998). Irwin (1998) suggested that for plane stress the energy based parameter G is related to the fracture toughness (Eq. (8.4)) as

$$G = \frac{K_1^2}{E} + \frac{K_2^2}{E} + \frac{K_3^2(1+\nu)}{E} \tag{8.23}$$

where the subscript of K identifies the mode of loading (I, II, or III) and ν is Poisson's ratio. GDS composites generally fail during loading mode I, and for these conditions

$$G = \frac{K_1^2}{E}$$

Failure is considered to occur when G reaches a critical value (GC); the failure criterion can be written as

$$G_c = \pi\sigma_f^2 c/E \tag{8.24}$$

It is essential to understand the microstructural mechanisms that are dominant during the failure of GDS composites, and further, how the toughness of the material relates to the microstructure. The toughening mechanisms of ceramics have been classified into three groups by Green (1998): crack–tip interaction, crack tip shielding, and crack bridging. The crack tip interaction (crack front bowing) and crack bridging are believed to be the major toughening mechanisms for GDS composites. It is, however, difficult to theoretically analyze the crack bowing mechanism because the properties of the interparticle bonding area are unknown. On the other hand, the effect of crack bridging on fracture toughness may be evaluated on the basis of Green's work (1998). Assuming that the mechanical properties at the interface in the mechanically locked areas are much lower than in the vicinity of ductile bridges, one can calculate the energy parameter G_C (after Green 1998) as

$$G_C = \chi V_V R \sigma_S \tag{8.25}$$

Here V is the volume fraction of bridges, $2R$ is the bridge width, σ_S is the yield stress of the ductile bridge area, and χ is a "work-of-rupture" parameter. The parameter χ varies considerably and, thus, must be accounted for. This parameter is defined by the bridge ductility and the amount of bonding bridge failures. If the bridges remain bonded, χ is approximated by unity. For partially debonded bridges, χ may approach a value of 8 (Green 1998).

In the case of GDS composites, one can assume that the volume fraction of bonded bridges is equal to the volume fraction of adiabatic shear bands $V_V = \Theta_{ASB}$. This result can be verified using Eqs. (3.8) and (3.10). Further, by using Eqs. (8.24) and (8.25), it can be shown that

$$\Theta_{ASB} = \frac{\pi \sigma_f^2 c}{\chi R E \sigma_S} \quad (8.26)$$

It has been demonstrated in the case of LPGDS technologies that the main microstructural characteristics of MMC are the mean free interparticle distance λ and the composite density ρ, the latter of which may be used to define the porosity $(1-\rho)$. By analogy, we may assume that the mean free interpore distance may be sufficiently defined by Eq. (8.2). λ_{pore} is determined by the number of pore intercepts per unit length of test line, N_{pore} and the volume fraction of pores , θ, as

$$\theta = \frac{1-\theta}{N_{pore}} \quad (8.27)$$

The experimental results of composite microstructure characterization (Fig. 8.21) demonstrate that the values of λ and λ_{pore} do not differ considerably with λ_{pore}, being principally dependent on the technological parameters of LPGDS. Therefore, the analysis of fracture characteristics is primarily based on the interparticle distance.

8.3.4.1 Analysis of LPGDS Composite Fracture Characteristics

Indeed, the failure criterions of GDS composites are dependent on the content of reinforcement particles and porosity. It is useful to discuss this issue with respect to how varying samples react to different tests (Fig. 8.2). GDS composites are pressure-dependent materials and for this reason loading schemes greatly influence their mechanical behavior. Thus, by determining the yield stress $\sigma_{0.005}$ (by uniaxial compression), fracture stress, σ_f, (by diametrical compression), and fracture stress of notched samples, σ_V (by diametrical compression of notched samples), we may evaluate interparticle bonding and deduce its dependence on both the microstructural characteristics of composites and the main LPGDS technology parameter, air temperature.

The influence of LPGDS air temperature on the properties of Al–30%Al_2O_3 composites with alumina particles of 10 and 45 μm is shown in Fig. 8.22. Re-

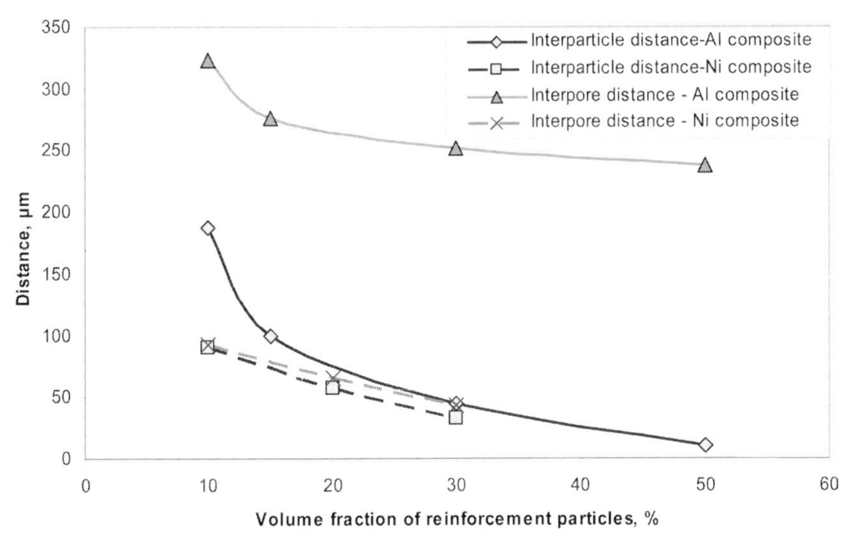

Fig. 8.21 Interparticle and interpore distance vs. content of reinforcement particles.

inforcement with larger particles leads to an increase in both the yield and fracture strengths of composites. Moreover, the ratio $\sigma_f/\sigma_{0.005}$ is significantly increased. It was discussed in Chapter 2 that an increase of temperature results in increased particle velocities and, consequently, kinetic energy (as does an increase in particle size). For this reason, interparticle bonding is more effective in this case than with smaller particles. The size of reinforcement particles also seems to be very important, in particular for LPGDS. This is because larger particles more readily compact the soft metal particles, as shown in Fig. 3.10. The effect of air temperature on σ_f is great, increasing it by as many as 2–2.5 times.

As shown above, the mean interparticle distance λ determines the strengthening of particle-reinforced composites. The relationship between yield stress $\sigma_{0.005}$, fracture stress of notched samples $\sigma_{f\,\mathrm{notch}}$, and λ (Fig. 8.23) demonstrates the strong dependence of these parameters on λ. If the yield stress decreases with an increase of λ, the $\sigma_{f\,\mathrm{notch}}$ increases for Ni–TiC composite, resulting in a maximum in the $\sigma_{f\,\mathrm{notch}}$ curve for the Al–Al_2O_3 composite. This indicates that there is only a specific range of interparticle distances that brings about a maximum $\sigma_{f\,\mathrm{notch}}$. For Al this interparticle distance is approximately 25–100 µm. Though it cannot be seen on the diagram for Ni–TiC, it is possible that a maximum in $\sigma_{f\,\mathrm{notch}}$ may occur at larger interparticle distances.

A comparison of the properties of composites made by various processing routes is crucial both for determining potential applications for LPGDS and improving the properties of LPGDS through postprocessing operations. Here classical powder metallurgy (PM), compaction sintering, LPGDS, and LPGDS

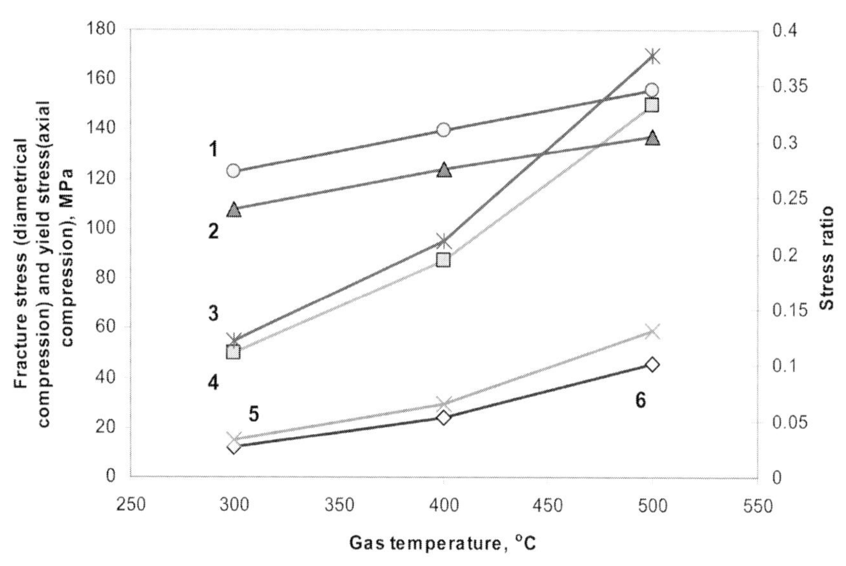

Fig. 8.22 Yield strength and fracture stress of Al–30%Al_2O_3 composites. (1), (3), and (5) stress ratio $\sigma_f/\sigma_{0.005}$, yield stress $\sigma_{0.005}$ and fracture tensile stress sf of Al–30%Al_2O_3 composite with reinforced particles of 45 μm size; (2), (4), and (6) stress ratio $\sigma_f/\sigma_{0.005}$, yield stress $\sigma_{0.005}$ and fracture tensile stress σ_f of Al–30%Al_2O_3 composite with reinforced particles of 10 μm size.

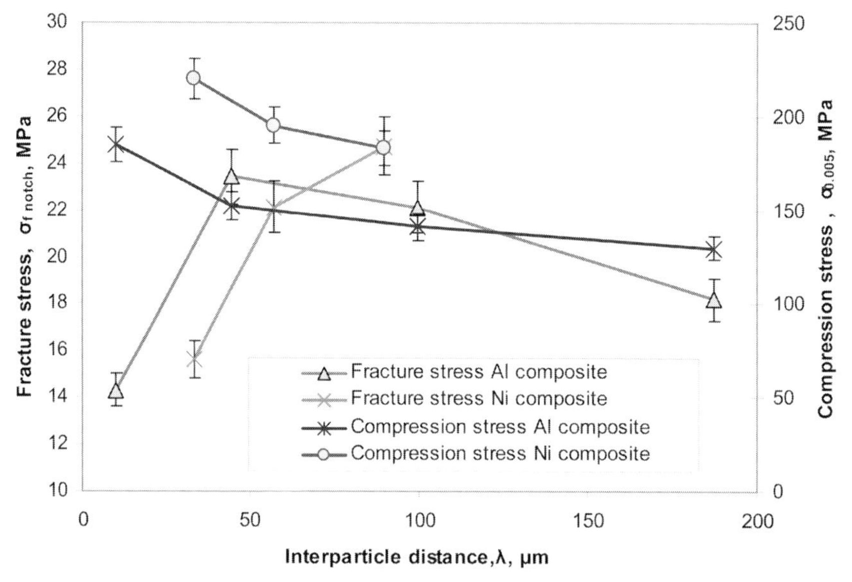

Fig. 8.23 Axial compression flow stress and diametrical compression fracture strength sf notch of the notched samples vs. interparticle distance.

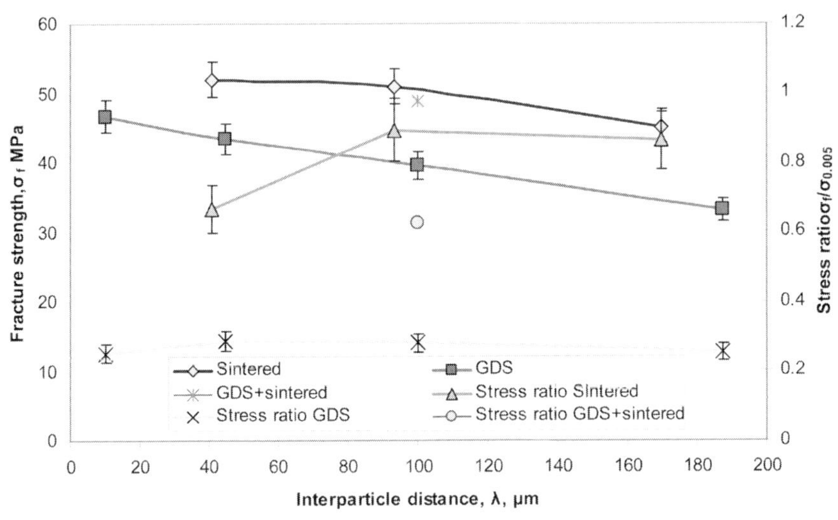

Fig. 8.24 Effect of various processing routes on fracture tensile stress s_f and stress ratio $\sigma_f/\sigma_{0.005}$ of Al–Al_2O_3 composites.

sintering are selected for comparison in Fig. 8.24. Experimental data reveal the positive influence of the sintering process (550°C, 1 h) on interparticle bonding for either compacted or GDS formed samples. In any case, the postprocessing of the samples leads to a three-fold increase in the ratio ratio $\sigma_f/\sigma_{0.005}$.

The fracture toughness K_{1C} (as defined by Eq. (8.4a)) for the diametrical compression of V-notched samples of LPGDS Al- and Ni-based composites are shown in Fig. 8.25. The data for V-notched samples correlate well with the previously published data of Clobes and Green (2002) for the fracture toughness of porous alumina. To ensure that the stress concentrations at the bottom of the notch are large enough to make the groove act as a sharp crack, the V-notch has to be sufficiently sharp with the notch-tip radius quite small (< 30 μm). The V-notches in the authors' tests consists of tip radii of about 10–30 μm. The fracture toughness of the Al- and Ni-based composites did not vary with the notch depth and thickness of the samples. The values of K_{1C} were within the range of 1.5–2.2 MPa $m^{1/2}$.

Finally, the volume fraction of adiabatic shear band areas should be determined. According to Eq. (8.26), this value should be equal to that of the volume fraction of bonded bridges. The results of this evaluation are presented in Fig. 8.26. The data reveal low volume fraction of bonded bridges for both Al- and Ni-based composites. More intensive deformation of particles in the Al–Al_2O_3 composites resulted in a higher volume fraction of bonded bridges than for the Ni–TiC composites. Thus, as suggested earlier, the values of Θ correlate with the porosity.

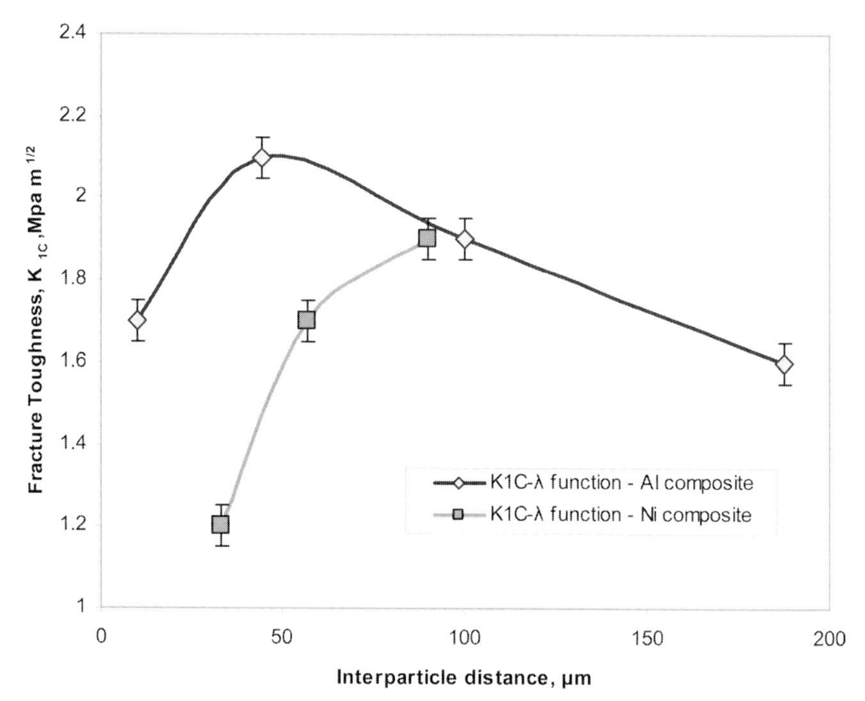

Fig. 8.25 Fracture toughness of the notched samples vs. interparticle distance.

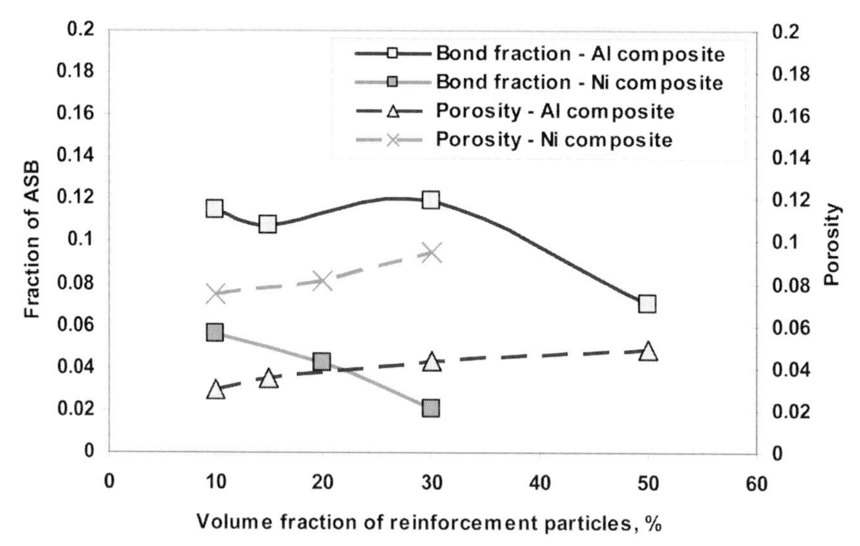

Fig. 8.26 Evaluation of volume fraction of bonded bridges with relationship (8.26).

8.4 Effect of Substrate Properties and Surface on the Deposition Process

8.4.1 General Analysis and Effects of Residual Stresses

The bonding of metals to various substrates by GDS has not received much attention until now. However, as these metals are able to bond in the solid state at relatively low temperatures they are gaining great interest for industrial applications. In any bonding process, the primary condition is that intimate contact is established between the materials, enabling interaction. In solid-state bonding this is achieved, at least partly, by exerting pressure at the interface. Usually, to achieve a rapid rate of interaction, the temperature should be as high as possible allowing the bonding to be completed in a reasonable time. Further, the substrate surface must be free of oxide and the other surface contaminants. The GDS process is characterized by high velocity particle impact which in turn leads to crater and adiabatic shear band formation at the interface. The interface temperature change ranges from 600 to 1200°C within approximately 10–40 ns. Despite this relatively short time interval of interaction and subsequent cooling, there is still some degree of oxidation at the interface. Thus, the use of controlled atmospheres is considered to be of great importance. In GDS processes in general, however, it is inherently difficult to control the oxidation process of the substrate. In fact, the only way to ensure a clean substrate surface is to destroy the oxide film by grit blasting and plastic deformation of the metal during the GDS process. Hence, the chief parameters in the bonding process are: particle velocity, temperature and impact, and cooling time of the bonded area.

Some data pertaining to the adhesion strength of HPGDS and LPGDS coatings are summarized in Table 8.4. Here the adhesion of Al, Cu, Ti, and Ni coatings on Al, steel and other substrates are of primary interest.

Currently, numerous techniques surrounding the thermal spray of coatings on Al substrates are being intensively developed. Aluminum alloys exhibit very advantageous mechanical properties, especially with respect to their strength-to-mass ratio. These days numerous machine parts are made of aluminum with many of them being coated by thermal sprays. However, spraying onto parts made of aluminum alloys is much more difficult than onto more traditional steel elements. In particular, the bond strength between most coatings and aluminum is relatively weak. The main reason for this is the high thermal expansion and conductivity associated with the rapid aluminum oxidation in the air. As a result a permanent, chemically continuous oxide barrier develops on the substrate. This oxide hinders close contact between the sprayed coating and the substrate. An oxide layer of about 0.7–1.0 μm is

Tab. 8.4 Adhesion strength and the main technology parameters of HPGDS and LPGDS processes.

Source	Coating composition	Substrate material	High pressure GDS					
			Helium			Nitrogen		
			Temperature, (°C)	Gas pressure (bar)	Adhesion strength (MPa)	Temperature, (°C)	Gas pressure (bar)	Adhesion strength (MPa)
Van Steenkiste et al. (2002)	Al	Al				250	20	35.9
Fakanuma et al. (2004)	Cu	Al	200	10–25	35	250	25	22.5
	Cu	Cu	200	10–25	45	250	25	15
Fakanuma et al. (2004)	Ti	Al	200	30	45–55			
	Ti	SS	200	30	45–60	250	30	18
Marocco et al. (2005)	Ti	SS				200	29	20
			Low-pressure GDS					
			Nitrogen			Air		
Authors	Al	Al	300	6	12.5–14	300	10	10–12
	Al–15%Al_2O_3	Al	300	6	16–18	300	10	15–18
Authors	Cu–10%Zn+15%TiC	Steel	400	6	32-35	400	6	25–30
Authors	Ni–20%TiC	SS	600	10	16–20	600	10	12–14
	Ni–20%TiC	Inconel	600	10	22–30	600	10	20–23

formed on the aluminum surface. Now, although grit blasting is effective at removing the oxide layer, its results are only temporarily.

One should also note the low adhesion strengths of thermally sprayed coatings on Al substrates (15 MPa). By contrast, detonation spraying on aluminum substrates only generates coating adhesions as low as 20 MPa. As shown by Bialucki and Kozerski (2006) for plasma spray, preheating Al substrates before spraying leads to a critical enhancement in the durability and strength of the bond between the layer and the Al substrate. Effective preheating temperatures often exceed 130°C.

The examination of HPGDS Al coating adhesion on an Al substrate made by van Steenkiste et al. (2002) reveals a higher value of adhesion strength as compared with other thermal spraying processes, about 30–35 MPa (see Table 8.4). This result seems to be a consequence of a relatively low temperature of deposition which results in both low thermal expansion of the substrate and low residual stresses in both the coating and the substrate.

It is generally acknowledged that the residual stress state in a thermal sprayed coating has two primary origins. During the first stage of deposition, individual molten particles heat the substrate and solidify. Complete contraction is not possible, owing to the presence of the substrate and/or the neighboring particles. This phenomenon leads to residual stresses, called "quenching stresses." The second stage of the spraying process is related to cooling of the coating. The presence of the cooling stresses is due to both the mismatch between the thermal expansion coefficients and the temperature difference between the coating and the substrate. Better adhesion between a coating and its substrate is also expected when the mean residual stresses in the region of the interface are as low as possible (Araujo et al. 2005). The data of Sampath et al. (2004) for Ni–5%Al coatings on a steel substrate reveal that residual stresses in GDS coatings are near to zero, indicating high potential for its effective application.

8.4.2 Microstructure Analysis of Interface

The properties and characteristics of the substrate surface are shown in Table 8.4 and Fig. 8.27. The ratio between the hardness of the substrate and spraying powder has a strong influence on the adhesion strength of coatings. The quality of adhesion between plasma-sprayed coatings and the substrate has been assessed by means of a scanning electron microscope. During examination, fractured surfaces of the coating–substrate system were observed at the side of the strayed layer.

Optical and SEM micrographs at the coating–substrate interface of samples studied are shown in Figs. 8.28 and 8.33. Examination of microstructure of Ni–

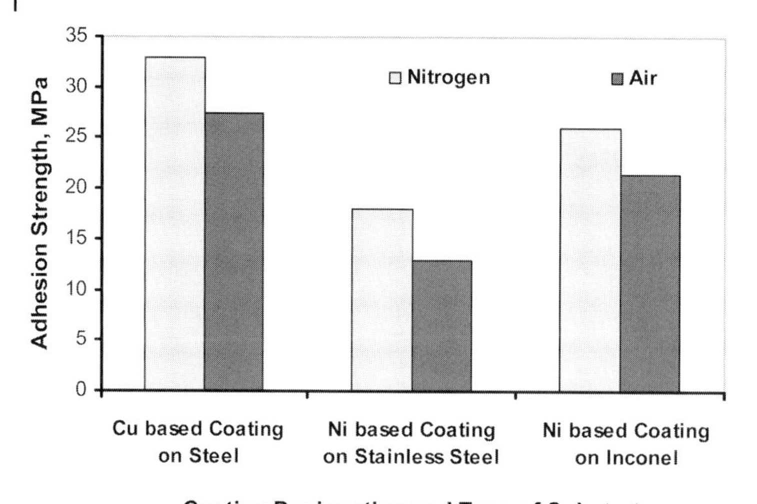

Fig. 8.27 Adhesion strength of LPGDS coatings.

10%TiC coating–Inconel interface, carried out by optical microscopes (Figs. 8.28(a) and (b)) reveals good adhesion at a well-defined coating–substrate interface. Micrographs made with a magnification of 200× show that the coating and substrate interpenetrate in the rough surface zone. Exceptional purity of the surface contact promotes the generation of microwelds between coating and substrate, which, in turn, enhances adhesion. However, one should see in the SEM of the fractured surface that the adhesion of TiC particles to the Ni matrix is poor (Fig. 8.29). Also, some Al oxide particles are seen at the interface. These particles seem to have penetrated into the substrate during sandblasting operation, diminishing the overall adhesion strength.

The microstructure observations of the Al–10%Al_2O_3 coating–steel interface (Fig. 8.30(a)) reveal numerous small Al_2O_3 particles at the interface. Further analysis of etched microstructure (Fig. 8.30(b)) shows numerous small grains at the interface, suggesting the possibility of a recrystallization processes. Both the small grain size occurring at the interface and the high content of Al oxide particles promote better adhesion. Besides the SEM observations, Al–10%Al_2O_3 layer fracture reveals regions of bonding between Al and Al_2O_3 particles (Fig. 8.31). Thus, as can be seen from the micrographs, strong local adhesion is characterized by both close surface contact and increased surface roughness. It can be concluded from Fig. 8.31 that Al oxide particles were embedded into severely deformed Al particles, ensuring effective particle shock consolidation.

Similar structures were observed in the case of Cu–20%TiC coatings. The microstructure of a Cu–20%TiC coating–steel interface is shown on Fig. 3.32(a). The results of the investigation show a possible benefit of recrys-

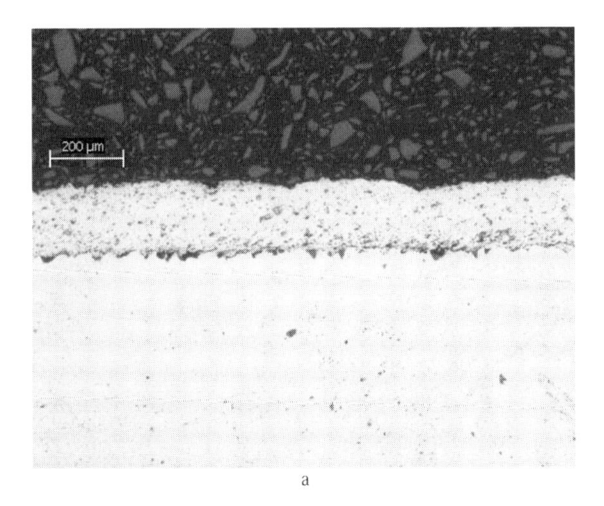

a

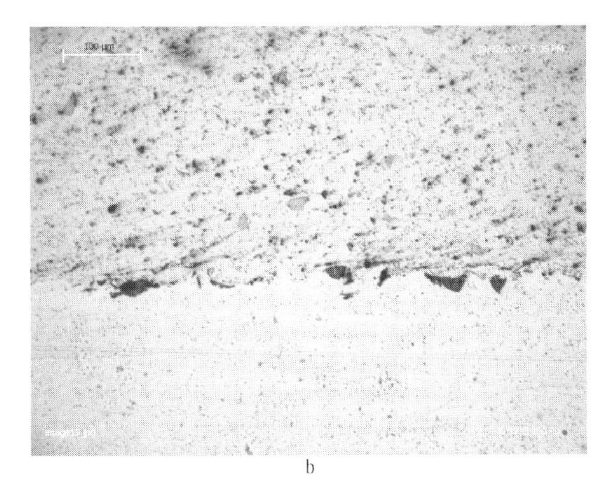

b

Fig. 8.28 Microstructure of Ni–10%TiC coating–Inconel interface.

tallization in the layers surrounding the interface (Fig. 8.32(b)) during the coating deposition process. Consequently, adhesion in all tested samples was determined to be about 30–40 MPa. SEM of a Cu–20%TiC coating-steel fractured interface after adhesion test revealed the areas of bonding between Cu and TiC particles (Fig. 8.33).

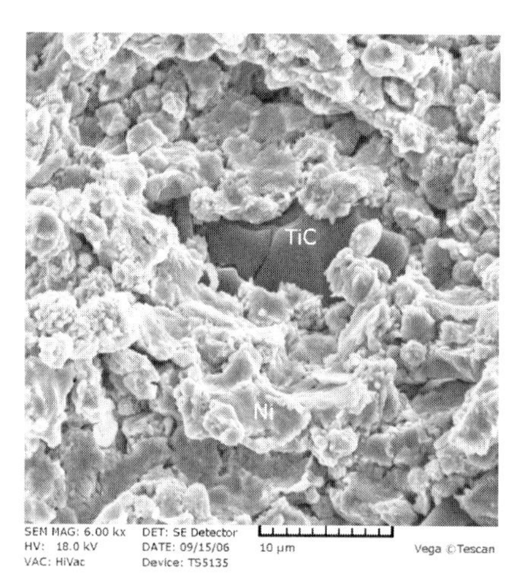

Fig. 8.29 SEM of an Ni–10%TiC coating–Inconel fractured interface after adhesion test. Weak bonding between the Ni matrix and the TiC particles can be observed.

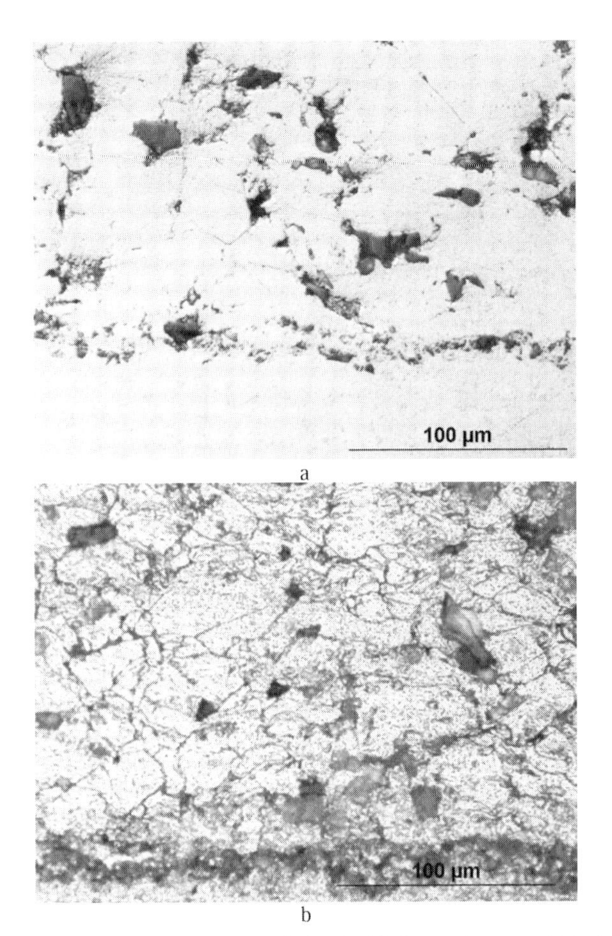

Fig. 8.30 Microstructure of Al–10%Al_2O_3 coating–steel interface. (a) Nonetched; (b) etched.

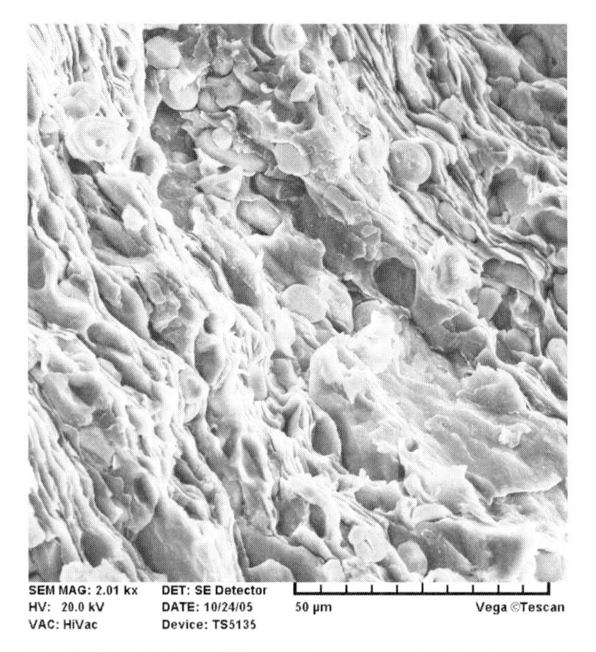

Fig. 8.31 SEM of Al–20%Al_2O_3 coating–steel fractured interface after adhesion test. The areas of bonding between Al and Al_2O_3 particles can be observed.

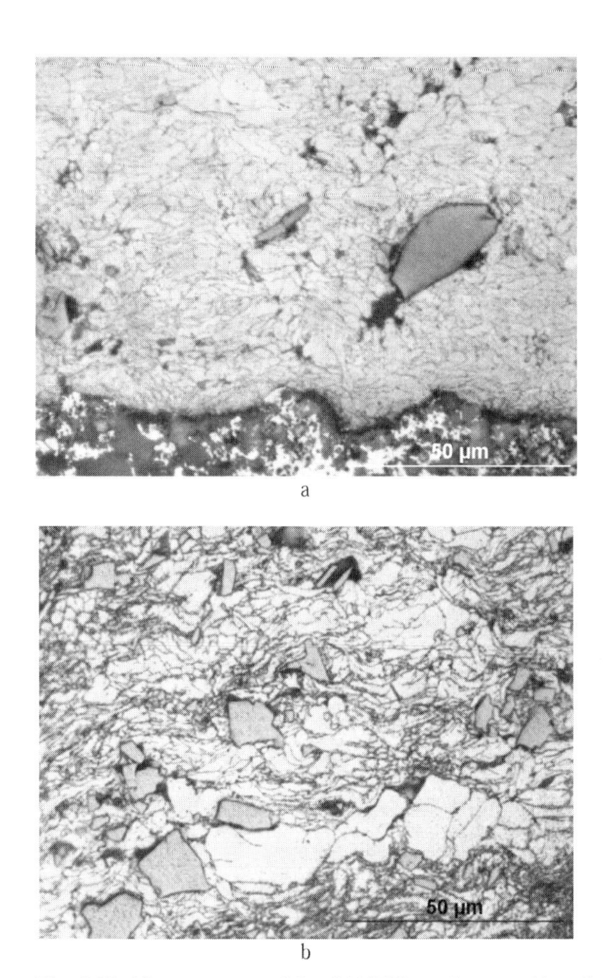

Fig. 8.32 Microstructure of Cu–20%TiC coating–steel interface (etched). (a) Interface; (b) an area near the interface.

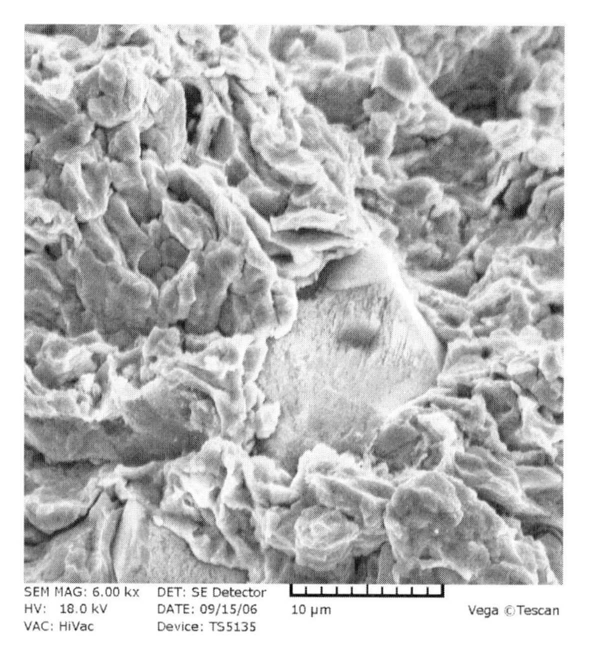

Fig. 8.33 SEM of Cu–20%TiC coating–steel fractured interface after adhesion test. The areas of bonding between Cu and TiC particles can be observed.

9
Low-Pressure GDS Applications

9.1
General Analysis

As shown in Chapter 1, advanced thermal spray technologies such as HVOF, wire arc or flame spraying, plasma spraying, and detonation spraying are widely used within the automotive industry, among many others. As an example, according to Barbezat (2006), a large portion of engine and transmission components are coated using various processes. While some of these applications have already reached maturity, many others are relatively new, representing an important market for the future of spraying technologies and, in particular, GDS. The range of materials that may be coated, as well as the quality of GDS deposits, depends to a large extent on the particle velocities that may be attained, as well as the mechanical characteristics of both the particles and substrate. From this viewpoint, it is beneficial to compare the capabilities of HPGDS and LPGDS processes in the context of their industrial applications.

As discussed by Villafuerte (2005), the application of helium as a processing gas for HPGDS provides the highest possible supersonic flow velocities. This technology, then, is capable of successfully depositing the widest range of materials in an oxygen-free environment (Lima et al. 2002). Unfortunately, helium is both relatively expensive and requires the use of special gas-handling and recovery/recycling equipment. Nitrogen, on the other hand, is considerably less expensive while being equally capable of providing an oxygen-free environment. The supersonic velocities obtained with nitrogen, however, are lower than with helium, limiting the range of coatings that can be produced. Air in fact offers the least expensive alternative but has the narrowest range of known applications, being generally limited to those processes that are not extremely sensitive to the presence of oxygen at low temperatures.

In the case of LPGDS, the particle velocities are in the range of 400–600 m/s which is only sufficient for the successful bonding of soft metals such as Zn, Al, and Sn. It has been shown in previous chapters that localized plastic deformation at the particle–substrate interface is necessary for the kinetic energy transformation. For this reason, successfully cold sprayed powders and sub-

Introduction to Low Pressure Gas Dynamic Spray Roman Gr. Maev and Volf Leshchynsky

ISBN: 978-3-527-40659-3

strates consist primarily of metals with relatively high plasticity. Nevertheless, as discussed in Chapters 3 and 5, there exist clear-cut distinctions between the behavior of single and collective particle groups when impacting a surface. It is for this reason that it is indeed possible to successfully spray various powder mixtures onto both metallic and ceramic substrates alike. In fact, the presence of a metal–ceramic powder mixture allows sufficient particle deformation to take place at particle velocities that are significantly lower than that required for the bonding of pure metals. Therefore, the range of materials that can be successfully sprayed by LPGDS may be expanded to include not only pure metals such as Al, Zn, Cu, Fe, Ni, and Ti, but also composites such as Al–Al_2O_3, Al–B_4C, Al–Zn–Al_2O_3, Cu–Pb–Sn, Cu–W, Cu–Zn–TiC, Ni–SiC, among others.

In order to emphasize the benefits of LPGDS, some specific applications are described in the sections to follow. To facilitate this task it is perhaps useful to classify these GDS technologies into four groups according to their mechanical, thermal and electrical, corrosion and biocompatibility, and technological parameters. Table 9.1 summarizes known and possible applications for gas dynamic spray technology in accordance with this classification.

Tab. 9.1: The development of GDS applications.

Desired properties	Applications	Examples	GDS technology comments
Mechanical properties	Metal restoration and sealing	Engine blocks, castings, molds and dies, weld seams, autobody repair, HVAC equipment, refrigeration equipment, heat exchangers, cryogenic equipment	1. The application of soft metals (Al, Zn, Sn) which is preferable with LPGDS 2. Need of portable mobile LPGDS equipment

Desired properties	Applications	Examples	GDS technology comments
	Wear resistant, low-friction coatings	Die components, valves, piston rod guides with low-friction coefficient	1. The application of wear resistant compositions (WC–Co, NiCr alloys) is possible with HPGDS 2. Some components with hardness about 200HB may be deposited by LPGDS
	Friction coatings	Increase friction of rolls for papermaking	Composite friction coating may be made by LPGDS
	Antistick properties	Deposits impregnated with release agents such as PTFE or silicone	The soft metal-based coatings with high porosity is being made by LPGDS
Thermal and electrical properties	Thermal barriers	Aluminum piston heads, manifolds, discbrakes, aircraft engine components	Al–Ti-based coatings and composite coating is being made by HPGDS and LPGDS
	Thermal dissipation	Copper or aluminum coatings on heat sinks for microelectronics	The Cu- and Al-based coating is being made by HPGDS with high deposition efficiency
	Soldering priming	Microelectronics components and CPBs	The soldering Ag and Sn composite coating is being made by HPGDS and LPGDS
	Electrically conductive coatings	Copper or aluminum patches on metal, ceramic, or polymeric components (McCune 2003)	The Cu- and Al-based coating is being made by HPGDS with high deposition efficiency

Desired properties	Applications	Examples	GDS technology comments
	Dielectric coatings	Ceramic coatings for aerospace, automotive, and electronic packaging	The ceramic-based composite coating is being made by HPGDS
Corrosion properties and biocompatibility	Localized corrosion protection	Zinc or aluminum deposits on affected helms, weldments, or other joints in which the original protective layer on the base material has been affected by the manufacturing process	1. The application of soft metals (Al, Zn, Sn) which is preferable with LPGDS 2. Need of portable mobile LPGDS equipment
	Biomedical	Biocompatible/bioactive materials on orthopedic implants, protheses, dental implants. Porous coatings of these materials on load-bearing implant devices facilitate implant fixation and bone in growth, replacing the need for cements and screws.	GDS of hydroxyapatite-based coating allows to retain the nanostructure of hydroxyapatite as compared to HVOF or plasma straying
	Decorative coatings	Metal and ceramic components	The portable mobile LPGDS equipment is preferable

Desired properties	Applications	Examples	GDS technology comments
Technology modification	Rapid prototyping and near-net manufacturing	Cold spraying can produce a well-defined footprint. Fabrication of parts with custom metal–matrix composite structure or gradient structures.	GDS allow to create the more flexible and cheaper technology for thin wall components.

9.2 Repair Applications of GDS Technology

Various surface engineering technologies such as laser cladding, HVOF, and plasma spraying are presently used to repair aerospace, automotive, marine, rail, or other such components where excessive wear has occurred. Here, GDS processes are utilized to reduce residual material stresses without perturbing the nanostructure of the buildup layer with excessively high temperatures. Also, these techniques are beneficial for repairing irreplaceable, high-value or over-machined components. The ultimate application of GDS, however, is to build components up from nothing; this may be accomplished using a laser cladding system and a 3D CAD drawing of the component. It is for this reason that GDS is emerging as a special technique for rapid prototyping (RP). Indeed, the combination of GDS with RP is an interesting approach to repair technology which offers an alternative to traditional tool builders, remanufacturers, and repair industries. For example, a LPGDS build up process may be readily implemented into the industrial workplace by means of a portable SST machine (Fig. 4.6), providing a cost cutting repair technology that may be employed to increase the wear resistance of mechanical components, in casting repair and sealing, and car body shape restorations. The repair of aluminum structures by GDS techniques is also of particular interest in reclamation applications. Instead of replacing an entire structure, it is sufficient in many cases to simply employ a surface coating. Hence tremendous savings may be realized through the application of GDS techniques (Djordjevic and Maev 2006).

9.2.1
LPGDS Composite Coatings for Mechanical Components

For further development to take place with respect to improving the wear resistance of mechanical components it is necessary to efficiently choose composite coatings for LPGDS that bring about the desires mechanical properties. Because of their versatility, then, metal–ceramic composite coatings are of great interest. For example, the incorporation of metal and ceramic particles within various alloys acts to increase their load-bearing capacity (Straffelini et al. 2004). This technology provides new opportunity for the employment of Al-, Ni-, and Cu-based metal–matrix composites (MMC) in applications where sliding resistance is of concern.

Some investigations have analyzed the wear behavior of MMCs for temperature ranges up to 200–500°C (Straffelini et al. 2004). It has been found that due to the thermal softening of the composite wear increases as temperature is increased, becoming significantly amplified above a critical temperature. The friction coefficient is also observed to increase as a result of the increase in the adhesion forces. Most experimental investigations in this study were carried out using hard steel as a counterface.

Of particular interest in the automotive industry is the investigation of friction and wear between MMCs and steel. This is particularly due to the possibility that exists in using these materials for on-line repairs of stamping dies. With respect to traditional cast iron, MMC GDS coating technologies also offer promising advantages with respect to in increasing the wear resistance of cast iron dies.

It is essential, then, that Ni- and Cu-based coatings containing W, Zn, and TiC reinforcement particles receive further investigation to determine their wear behavior, which may be different depending on the counterpart material used. Straffelini et al. (2004), for example, investigated the friction and wear behavior of two Al–MMCs against both organic and semimetallic materials. Wear rates were found to be lower with respect to organic materials than against semimetallic materials. This is due to the formation of a graphite-rich transfer layer in the organic materials; whereas, the semimetallic materials contain metallic chips or fibers that are highly abrasive. Understanding the nature of the tribological contacts is very important when attempting to explain the observed friction and wear behavior in blank/die couplings. In a real cast iron die component surface, the nominal contact pressure, p_0, typically varies between 1 and 100 MPa, and the sliding velocity, v, between 1 and 5 m/s. Most laboratory investigations regarding the friction and wear behavior of materials for die applications are carried out using ball-on-disc and block-on-disc wear rigs. In the present investigation, ball-on-disc tests were employed using a procedure for increasing the load on the friction to study the wear behavior of both Ni- and Cu-based MMCs against a Si_3N_4 ball.

9.2.2
LPGDS Technology Characterization and Experimental Procedure

The spraying of Ni- and Cu-based powder mixtures on steel and cast iron substrates was performed by a portable apparatus SST Centerline Windsor Ltd (Fig. 4.5) with the radial injection of powder into the divergent part of a supersonic nozzle (Fig. 1.1(b)) (Maev et al. 2004, Villafuerte 2005). A specialized equipment set was developed for this work by Centerline Supersonic Technology Division (SST) and is shown in Fig. 9.1. While the details of LPGDS technology were shown in Chapter 2, it is important to emphasis that the compressed nitrogen pressure was kept constant at 0.6 MPa, which is quite low, compared to other cold spray methods (usually in the range of 2–5 MPa) (Assadi et al. 2003, Van Steenkiste et al. 1999). Due to the fact that special high-pressure pumps are no longer needed, this feature makes GDS with powder radial injection technology much more suitable for industrial environments. Additionally, nitrogen consumption is lower than in typical cold spray systems, requiring only about 0.2–0.3 m^3/min (as compared to typical rates of 1.3 m^3/min) (Van Steenkiste et al. 1999). The particle velocities at the exit plane of the supersonic nozzle, as measured by a laser Doppler velocimeter, were in the range of 550 m/s. The powder feeding rate was varied in the range of 0.5–1.5 g/s, while the standoff distance from the exit of the nozzle to the substrate was held constant at 10 mm. A rectangular nozzle with an exit aperture of 3.5 mm $\times$10 mm was used. Commercially available nickel and copper powders (−50 μm), TiC powder (−45 μm), zinc powder (5–10 μm) and tungsten powder (3–10 μm) were used as the base materials. Cu (45 vol.%)–W (25 vol.%)–Zn (30 vol.%) and Ni (65 vol.%)–TiC (35 vol.%) powder mixtures were applied.

To define the structure and properties of the coatings, tribology tests, and metallography examinations were made. Tribological tests were performed by means of a customized ball-on-disc tribometer under dry sliding at room temperature (25°C). The relative humidity of the ambient atmosphere was 40%. The counterpart was a 3 mm diameter Si_3N_4 ball. In order to retain uniform test conditions, a new ball was used for each test and the diameter of a wear track on the disc was maintained constant between wear tests. Prior to performing friction and wear evaluation, the surfaces of the discs were polished and ultrasonically cleaned in acetone for 10 min. The constant sliding speed was 0.8 m/s. The friction coefficient was monitored via computer during the test, while the wear resistance of the coating was evaluated by examining the scar width. The wear tests were interrupted at fixed times to perform measurements. The measurements of the scar width were done at four representative times during the wear test. Three friction tests were performed for each condition and the results were reported as an average value. The effect that the test interruptions had on the deviation of the scar width was determined by comparing this data with data obtained from continuous wear tests

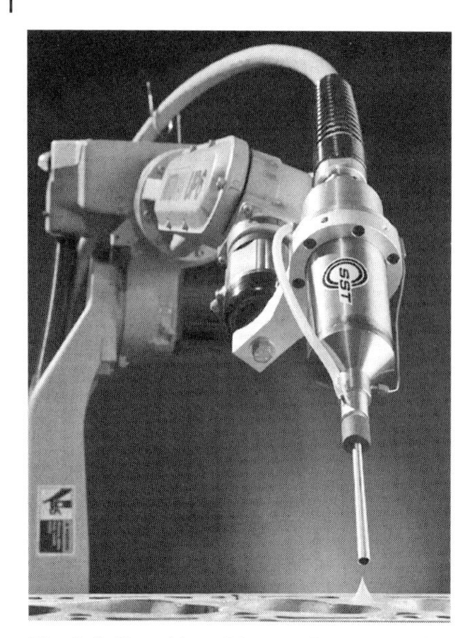

Fig. 9.1 Portable cold spray Centerline SST equipment.

taking place over the same period of time. This yielded a width deviation of no more than 10%. The microstructure and worn surfaces of the coatings were characterized using an optical and scanning electron microscope (SEM JSM-5800LV Jeol) coupled with an energy dispersion spectroscope (EDS System γ-PGT). The average chemical compositions of the coatings were measured by EDS at five different locations. Microhardness measurements were performed with a MHV-2000 digital Vickers microhardness tester using 0.05 N load and 15 s loading.

An alternative technique to evaluate wear resistance is the scratching test. This technique allows one to obtain wear resistance data for very small and confined regions – about 10–50 μm. For this reason the method provides a powerful tool with which to evaluate the mechanical performance of specific regions of the microstructure. Williams (1996) utilized this method to compare the scratch width of different phases in GDS coatings. The microscratch tests were performed in a modified microhardness tester with a Vickers square pyramid having a normal load of 0.01 N. An Olympus PMG3 optical microscope with an ISA4 image analysing system was used to measure the microscratch width and coating porosity.

9.2.3 Results and Discussion

9.2.3.1 Characterization

Figures 9.2 and 9.3 show optical micrographs of the polished and unetched cross-section of the Cu- and Ni-based coatings; typical features of GDS deposited composite structures are shown. Chemical analysis of the as-sprayed Cu-based coatings indicated a composition of Cu-25W-25Zn (at.%), and Ni–TiC coatings to be Ni–25TiC (at.%). Thus, the volume content of coating ingredients was Cu(45%)–W(28%)–Zn(27%), and Ni(62%)–TiC(38%). Comparison with the initial composition of powder mixtures reveals a small decrease in the content of Zn in Cu-based coatings and a decrease in the content of Ni in Ni-based coatings by 3%. This is perhaps due to the more intensive rebounding of Zn and Ni particles during deposition. In Cu-based coatings (seen in Fig. 4.2(a)), the black contrast corresponds to pores and fine oxide layers, the deeper gray contrast corresponds to W and W-based phases, and the lighter gray to Cu- and Zn-based phases. Table 9.2 shows a summary of the measured properties of the Cu- and Ni-based coatings on cast iron and steel substrates. Note both the relatively low porosity of these coatings and the higher hardness of the Ni–TiC coatings.

Tab. 9.2 Coating properties.

Coating	Adhesion strength (MPa)		Hardness HV0.1		Porosity (%)
	Iron G2500	Steel 1025	Iron G2500	Steel 1025	
Cu–W–Zn	32.5	33.7	160	158	2.1
Ni–TiC	25.1	25.7	220	223	3.5

9.2.3.2 Sliding Wear Behavior

Figure 9.4 shows variations of the friction coefficient of the coatings as a function of the sliding distance under a constant sliding speed of 0.8 m/s for different normal loads. All tests showed that the friction coefficients increased right after the start of the test, fell slightly, and then slowly rose again, eventually achieving a steady state. The difference in the magnitude of the friction coefficients among the coatings was inconsequential. The average ball-on-disc dry friction coefficients of the studied coatings were in the range of 0.65–0.7. Comparison of these values with the friction coefficients of uncoated substrates (Bateni et al. 2006, Terheci et al. 1995 reveals that they are slightly higher by 0.1–0.2. Figure 9.5 shows the wear scar width of Ni–TiC and Cu–W–Zn coatings, as well as cast iron substrate at a load of 2 N. The scar width for Ni–TiC

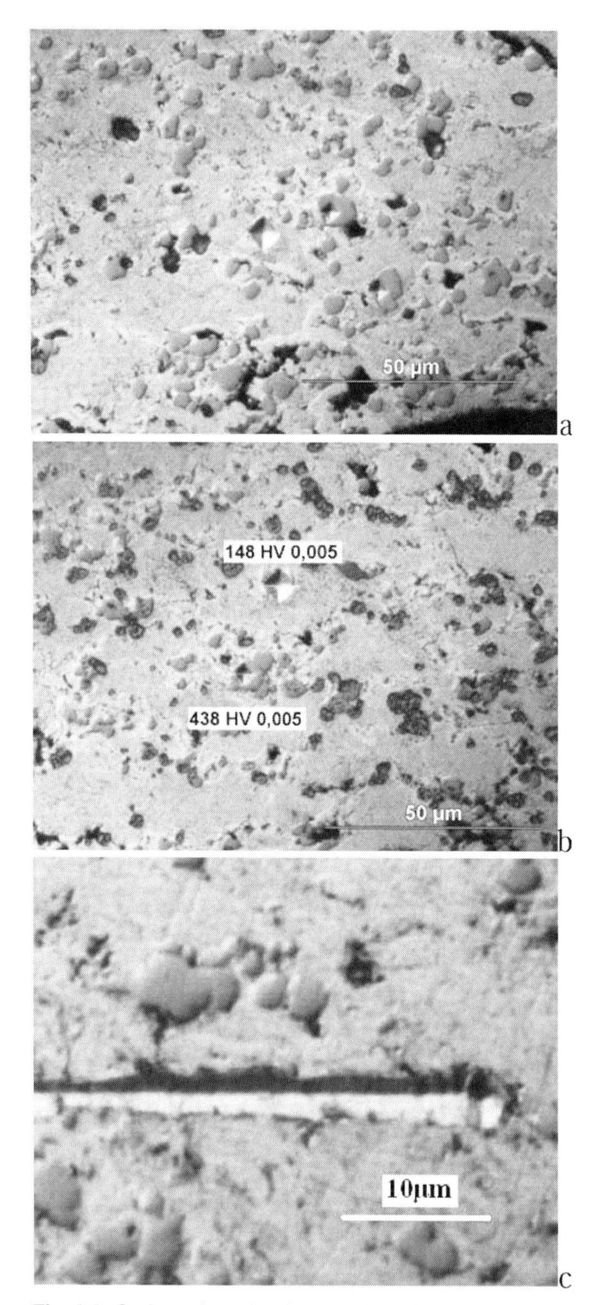

Fig. 9.2 Cu-based coating (composition Cu (50%)+W(25%)+Zn(25%)). (a) microstructure; (b) microhardness measurement results; (c) microscratch image.

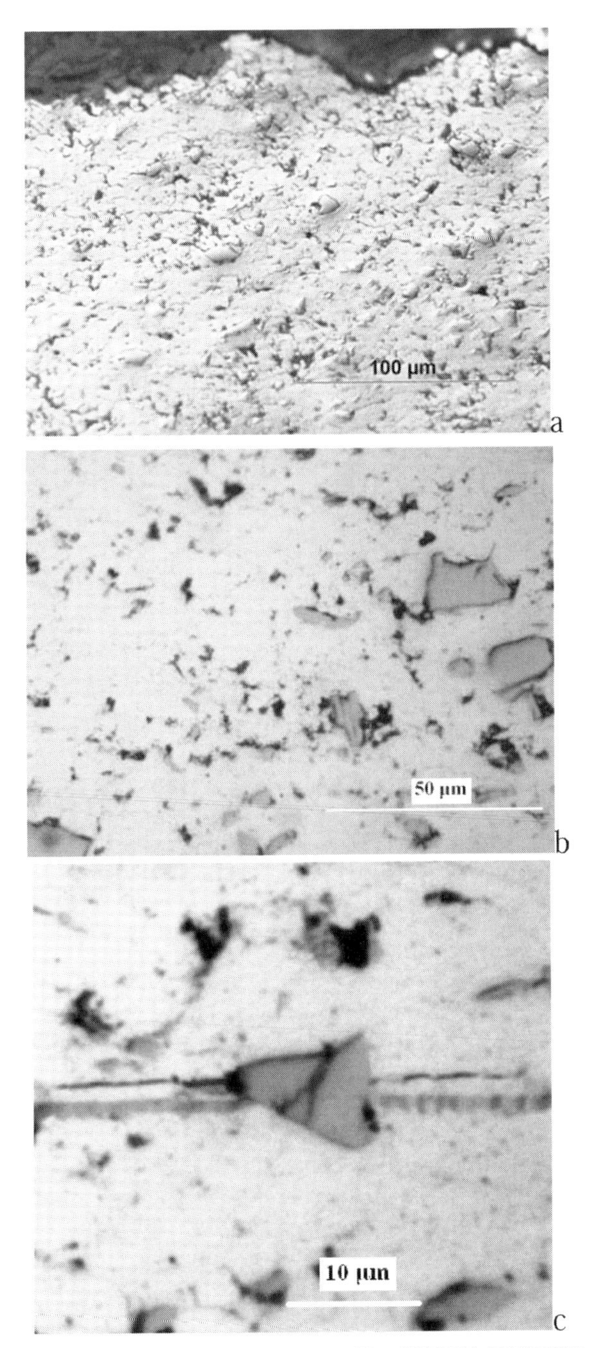

Fig. 9.3 Ni-based coating (composition Ni(75%)+TiC(25%)). (a) Microstructure; (b) microcracks near the TiC particles; (c) microscratch image.

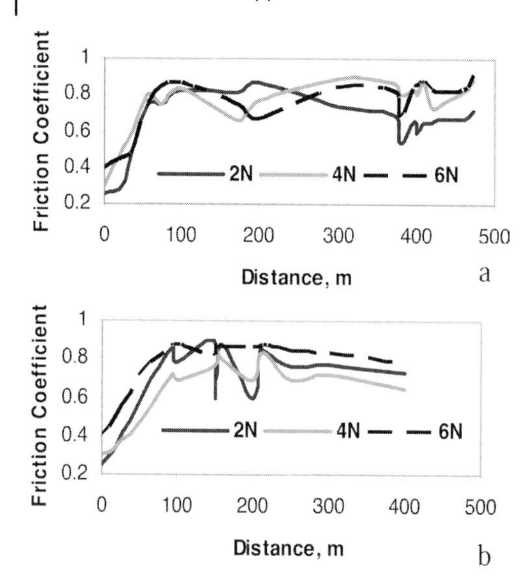

Fig. 9.4 Friction coefficient vs. sliding distance for (a) Ni–TiC coatings and (b) Cu–W–Zn coatings at various normal loads.

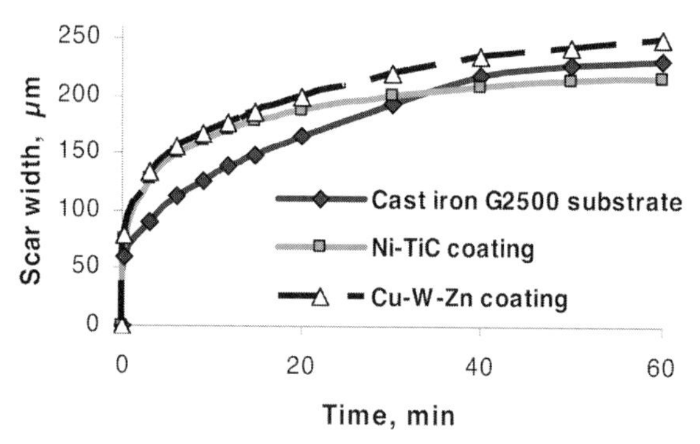

Fig. 9.5 Ball-on-disc wear test results for cast iron. Ni–TiC and Cu–W–Zn coatings at a normal load of 2 N.

composite coatings appears lower than for Cu–W–Zn coatings and the width appears to attain a plateau level over time. It is also apparent that the wear resistance of the coatings is close to that of the cast iron substrate.

9.2.3.3 **Analysis of Worn Surfaces**

An SEM image of the worn surface of a Cu–W–Zn coating after a ball-on-disc test is shown in Fig. 9.6. Here, areas of intensive wear particle generation and compaction are denoted with arrows. In order to determine the possible mechanisms of this wear particle generation, microscratch tests were executed. Figures 9.2(c) and 9.3(c) show the morphologies of the worn Cu–W–Zn

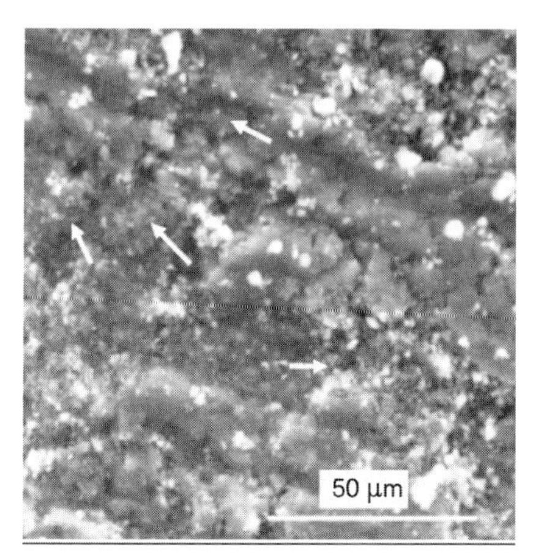

Fig. 9.6 SEM image of worn surface in the area of a wear groove on a Cu–W–Zn coating after a ball-on-disc test.

and Ni–TiC composite coatings after executing microscratch tests with a normal load of 0.01 N. A furrow of deformed material can be clearly observed on both coatings. This seems result from the microploughing action of the indenter, as well as from wear debris generation during the microscratch test. The worn surfaces can be divided into two regions with varying color: rough gray regions and smooth dark regions. For the Ni–TiC composite coating, TiC particles which would resist microploughing may be embedded into the worn surface (Fig. 9.3(c)). Also, a small volume of debris from the soft matrix can be observed on the worn surfaces (Fig. 9.2(c)), likely resulting from the sliding of the indenter. This indicates that plastic deformation most likely took place in the soft matrix. Despite the use of optical micrographs, the exact wear damage mechanism of the coatings remained unclear. However, it can be seen (Figs. 9.3(b) and (c) and 9.6) that microcracks initiated and propagating through the weakest boundaries between the particles, result in fractures which generated wear debris.

The friction coefficient depends primarily on the surface topography, material combinations and sliding conditions (contact mode, normal load, and sliding speed, etc.). The rapid initial increase at the beginning of the ball-on-disc test (Fig. 9.4) may be caused by high initial adhesive contact between the Si_3N_4 ball and the coating material. At this stage, the scar width of the coatings is bigger than that of the cast iron substrate. This is due to the presence of graphite, which seems to act as a lubricant at the sliding interface (Fig. 9.5). Also, compressive stresses on the coating during sliding lead to a slight decrease in the wear (after an initial increase), similar to the findings of Terheci et al. (1995). During the steady stage, debris is generated and compacted (Fig. 9.6), likely causing an increase in the frictional force and, in turn, increasing the friction coefficient. In both of the coatings that were examined, it is believed that the process of wear debris generation on the sliding interface reached a steady state. Also, the high-friction coefficient appeared to act as a stabilizer in a fashion similar to studies by other investigators (Xu et al. 2004, Suh 1986.

The sequence of mechanisms responsible for complex wear and friction phenomena may be summarized as follows (Mohanty and Smith 1996):

i. Particle detachment occurs, including adhesion, delamination, truncation, and ploughing.
ii. A transition from two to three body contacts takes place as a result of these wear mechanisms (formation of a "third body").
iii. Plastic and elastic deformation of the sliding bodies takes place due to debris breaking, rolling, and sliding.
iv. Wear track trapping with subsequent phenomena are observed such as ploughing or milling of the wear debris.

9.2.3.4 **Wear Microstructure**

During the particle detachment stage of ball-on-disc sliding wear tests, shear stresses resulting from the normal load and friction force on the contact surface promote soft matrix deformation and material removal. A comparison of microscratch tracks (Figs. 9.2(c) and 9.3(c)) demonstrates the effect of matrix microhardness on the scratch width. The copper matrix, with hardness HV = 140–145, exhibits scratch widths in the range of 3–4 μm (Fig. 9.2(c)), while the nickel matrix, with hardness HV = 300–350, has a scratch width of 2–2.5 μm (Fig. 9.3(c)). It should be noted that these coatings do show the same wear resistance at the start of sliding (Fig. 9.5).

9.2.3.5 **Wear Process**

It is suggested in this case that wear particle detachment at the first stage may be directly related to interparticle bond strengths (contrary to that of cast iron or steel substrates). In the following stages, the accumulation of the wear debris and strain hardening of the surface leads to a change in the mechanism of coating wear. The surface and subsurface of the coating are subjected to cyclic tensile and compression stresses (Xu, Zhu et al. 2004, Suh 1986), which may be referred to as a low-frequency fatigue process. This results in similar wear rates for both the coatings and iron substrate (the plateau on the Fig. 9.5). The adhesion strength of the coatings prepared by GDS is considerably lower than the casting material yield strength due to the impact features of the sprayed structure (see Chapter 3). However, the high content of defects such as porosity and microcracks (appearing primarily at the boundaries between the thin oxide sheets and the deformed particles) leads to relatively low-adhesion strength between the coating and the substrate. Voids and microcracks grow, propagate and coalesce, resulting in failures similar to previous observations (Bateni et al. 2006, Terheci and Manory 1995, Xu et al. 2004.

The higher wear resistance of Ni–TiC composite coatings may be attributed to hard TiC particles protruding from the Ni-rich matrix, which are able to sustain higher surface pressures. This effectively resists microploughing and microcutting by the Si_3N_4 ball. The microscratch test results of Ni–TiC composite coatings described above confirm this suggestion. It is indeed significant from an application perspective that LPGDS technology is able to repair cast iron die components with coatings having high resistance to scratching.

9.2.4
Casting Repair

In many industries such as automotive and aerospace, competition is becoming increasingly intense. As suppliers and manufacturers of components strive to maintain profit margins by reducing cost, it is evident that manufacturing processes must both become more productive and generate a higher quality product. In order to achieve the required mechanical specification and density, the casting process should be controlled and modified to ensure that the parts produced are reliably free from macrostructural defects such as incomplete filling or hotspots due to restricted flow (Evans et al. 1991).

In particular, due to ever-increasing standards and requirements, progress relating to automobile casting has become increasingly important. For the most part, this development is limited to the basic mechanical problems associated with castings such as leakages. These leakages are caused by several forms of fissuring. It is necessary, then to find a cost effective means by which to repair these defects without fundamentally altering the properties of the casting.

GDS coating technologies which integrate surface preparation, coating material selection, complete process control and finishing technology provide an innovative method for the salvage of castings and cast components. Over the past years, the economic nature and proven reliability of this repair method has begun to gain widespread acceptance throughout the industry. Documented and approved applications are being performed by casting plants, engine and equipment rebuilders, job shops, etc. It is in these applications where LPGDS technology, due to the portability of the SST machine, may be readily adapted to industrial environments.

Both Fe and Al castings are being deposited with Al- and Cu-based composite coatings. The choice of powder composition for a particular coating is governed by thermal expansion coefficients, adhesion strength, and the density (or porosity) of the coating–substrate interface, ultimately defining the

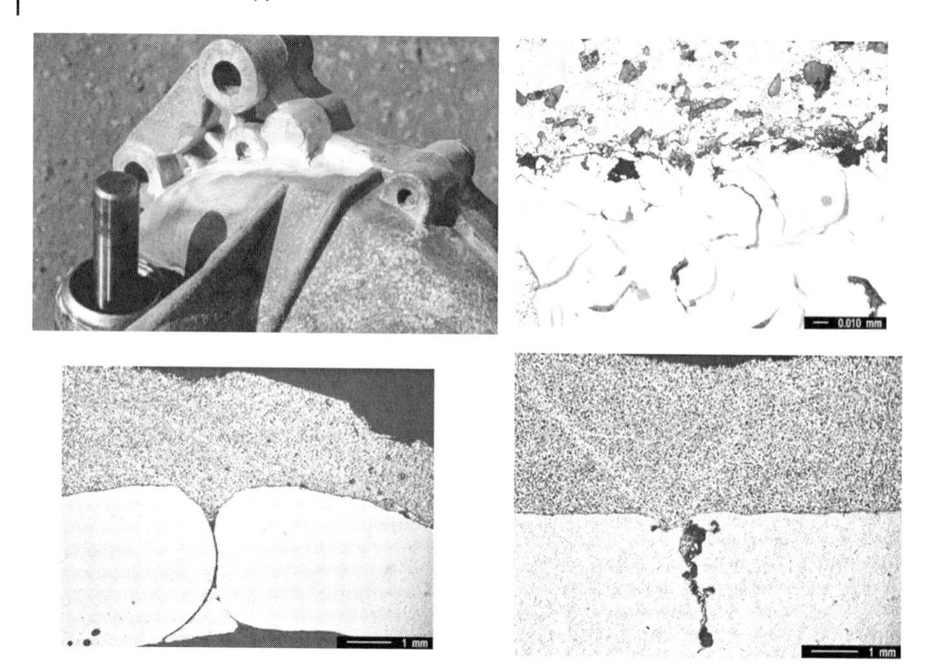

Fig. 9.7 Cast iron G2500 gear box repair. (a) General view; (b) microstructure; (c), and (d) macrostructure of repaired area.

sealing capability of joint. Results from the repair of an iron casting gear are shown in Fig. 9.7. The macrostructure of layers built up on the cast defects (Figs. 9.7(c) and (d)) reveals that the powder coating is very effective in filling flaws, voids, or other cast defects. The particles are trapped within a gap of 0.2–0.3 mm and stick to the substrate surface. The microstructure of both the coating and iron substrate is shown on Fig. 9.7(b). It can be seen that the coating–substrate interface is without cracks, and also, that some pores appear to be formed by Al_2O_3 particles which are removed during cutting and grinding. Other applications that have been met with success relating to the engine block, engine block head, and gear box are shown in Fig. 9.8. Here the necessary fatigue strength is achieved through the optimization of the constructive design and the appropriate selection of materials. This allows for adjustment in the thermal expansion coefficients of both the coating and substrate.

9.2.5
Casting Die Components Repair

Castings die components are common in automotive stamping. The metallic properties that are required for these particular applications can be attained through altering the raw material mix, resulting in a more robust component. For components which may be subjected to thrust or wear conditions, stability,

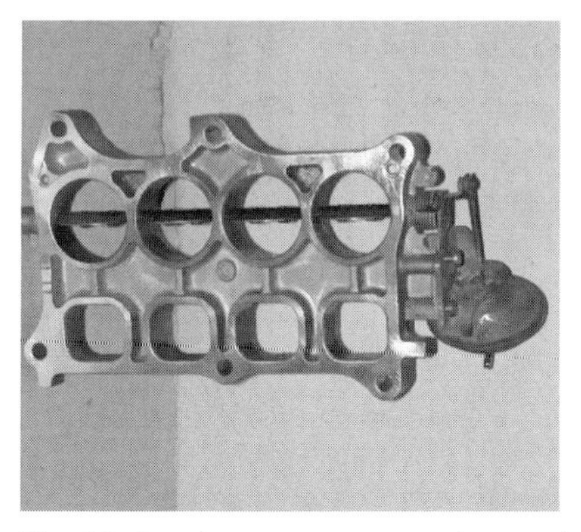

Fig. 9.8 Cast iron component repair examples.

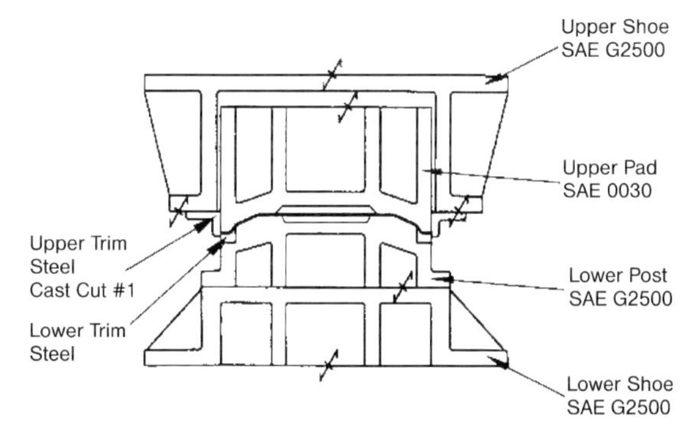

Fig. 9.9 Typical cast components of a die.

and compressive strength properties can also be added. The main advantage of a casting is that it is easier to cast a shape than to build it up using several pieces of steel. In a casting, steel can be poured into structurally strong components without the added weight of a solid steel construction. Castings are typically used in the main body construction of components such as the upper or lower die shoe, upper punch, lower post, binder, driver, cam slide, or risers (Fig. 9.9). In order to reduce machining time, castings with pockets and shoulders for tool steel, or wear plate inserts, are prepoured into shapes that require minimal surface removal. Items such as holes, pockets, and ribs, as well as lifting considerations may also be incorporated into the component design where necessary. Common casting materials include iron, steel, and tool steel.

Figure 9.7 demonstrates the opportunity for the application of LPGDS technology in the repair of die components (Beneteau et al. 2006). The properties of these coatings must be adjusted according to materials listed in Table 9.3.

Tab. 9.3 Construction materials and corresponding die components to be repaired construction.

Construction	Application	Material specification	
Casting	Inner panel die shoes adaptors, risers, drivers, cam slides	SAE-G2500	Grey cast iron, medium strength
Casting	Forming and restrike pads, draw rings, cap and post	SAE-G3500	Grey cast iron
Casting	Outer panel and inner door die shoe, lower ring, cap	SAE-0050	Cast steel
Boiler plate	Die shoes, bull-ring plate, special die bould-ups	SAE-1025	Machine steel
Composite	Scrap cutter straight cut trims	SAE-D2; SAE-1020	Tool steel; machine steel

9.2.6 Car Body Shape Repair

Due to its high efficiency, sheet metal forming is one of the most common metal manufacturing processes in industry. It is used to produce complex parts from flat sheet stocks covering a spectrum of products like car body components, shielding cans for electrical and housing equipment, etc. Strong demands for cost reduction, shortened development cycle, and higher quality require robust technologies which enable tool development, procedure and product optimization, final part damage restoration, and/or car body shape repair. LPGDS is the most suitable technology for this purpose because allows repair to take place either by hand or by means of a robotic arm The main powder compositions developed for this application are mixtures of Al and Zn, achieving a high density of coating with suitable adhesion strength. An example of an on-line car body shape repair procedure within an assembling plant is shown in Fig. 9.10.

9.3 Hardening by LPGDS Deposition

9.3.1 General Remarks

The unique features of GDS technologies have prompted much research regarding the optimization of this process (Papyrin 2001, Stoltenhoff et al. 2000).

a

b

Fig. 9.10 Automotive body restoration. (a) Defect; (b) after restoration by low-pressure GDS.

However, research to date has been concentrated on high-pressure GDS processes, where powder materials are injected along the axis of the supersonic nozzle. Low-pressure, radial-injected GDS provides an alternative to this technology. This spray technique is characterized by lower particle velocities and may be realized with a portable GDS machine (Maev et al. 2005). As stated before, the lower velocities associated with LPGDS may be overcome depositing powder mixtures and adjusting the spraying parameters.

It has been shown in Chapter 1 that GDS is considered to be a powder shock compaction and consolidation process (Stoltenhoff et al. 2000, 2001). The shock compression of powder mixtures produces a highly activated state by intimately dispersing the reactants, generating defects and introducing significant grain size reduction via the fracturing or formation of subgrain structures (Graham and Thadhani 1993). These effects result in significantly increased mass transport rates and enhanced chemical reactivity of powders, creating new paths for the motion of point defects along grain boundaries. The additional activation of shock powder consolidation through the application of GDS powder mixtures is believed to diminish the critical velocity limit. The aim of this section, then, is to demonstrate the possibility that exists to spray a Ni composite coating by low-pressure GDS. It is in fact possible to create wear resistant and corrosion resistant Ni coatings by means of an environmentally friendly technology. This would eliminate the need for galvanic techniques which currently dominate Ni composite processes. To accomplish this, the Ni powders should be reinforced by hard particles such as SiC, TiC, and others. Additional powder ingredients such as tungsten and Ni–Cr alloys may also be included to facilitate particle shock consolidation.

In a study by the authors, commercially available Nickel powder (−300 mesh), SiC powder (10–20 μm) and tungsten powder (5–20 μm) were used as base materials, creating two component Ni–SiC and Ni–W powder mix-

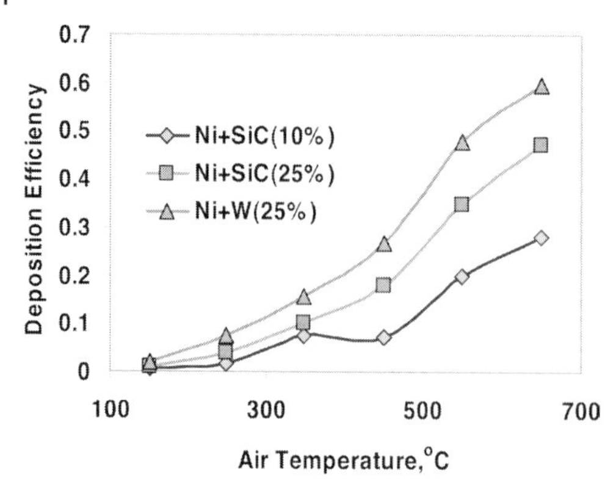

Fig. 9.11 Deposition efficiency as a function of air temperature for Ni–SiC and Ni–W powder mixtures (content in volume%).

tures. These materials were then sprayed on an Al substrate and the resultant coatings were examined. Further descriptions and results of this study are described below.

Characterization of the coatings was done through microhardness measurements and microscratch tests. The microscratch tests were performed on the coatings made of Ti (−270 mesh) and Inconel 600 (−325 mesh) powders on Al substrate to more carefully examine the microstructure features of a separate particles during the consolidation process. The microstructure of the coatings was examined using an optical microscope and SEM (JSM-5800LV Jeol) after completing standard metallographic preparations.

9.3.2 LPGDS of Ni–SiC Powder Mixtures

9.3.2.1 Deposition Efficiency

Experimental results determining the deposition efficiency (DE) are presented in Figs. 9.11 and 9.12. The experimental data reveal relatively low-deposition efficiency for the Ni-based powder mixtures at air temperatures up to 500°C. Particle velocity measurements yield values of 550–600 m/s in this temperature range. By comparison, Raletz et al. (2003) find the particle velocity for a high-density Ni coating to be approximately 807 m/s. Further increases in the air temperature enable higher deposition efficiencies for the studied mixtures.

The main result of GDS deposition efficiency analysis is the determination of the low-pressure GDS technology parameters, allowing Ni-based composite coatings to be generated at relatively low-particle velocities. This is achieved by adding SiC or tungsten particles. The data for adhesion strength, hardness, and density are shown in Table 9.4. Comparison of the coating den-

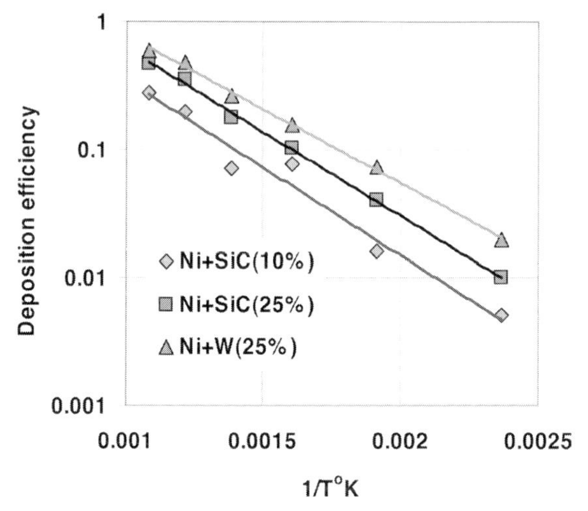

Fig. 9.12 Exponential approximations of the function DE $= f$ (1/Tair) for a Ni-based powder mixture.

sity reveals that an increase of the SiC content gives rise to increased porosity. This is due to the high concentration of pores at the boundaries between the Ni matrix and the SiC particles, as is illustrated in Fig. 9.13. Each interface between the SiC particle and Ni matrix contains pores. Many separate pores can be seen in the Ni matrix as well.

Tab. 9.4 Deposition efficiency, density, and mechanical properties of Ni-based composite coatings.

Coating	Deposition efficiency (T_{air} = 450°C) (%)	Adhesion strength (MPa)	Hardness (HV0.1)	Porosity (%)
Ni–SiC (10%)	6	20.5	280	6.83
Ni–SiC (25%)	15	25.1	330	7.5
Ni–W (25%)	28	40.2	300	5.34

Deposition of the Ni–W composition brings about considerable change in the mechanism of coating formation. One can see in the microstructures shown in Fig. 9.14 that the coating porosity is much less than that of Ni–SiC. The deformation of Ni particles appears to be more extensive, leading to the closing of pores in the Ni matrix. The more intensive deformation and densification of Ni particles seems to be due to the high-kinetic energy of the W particles, which collide with Ni. Also, the kinetic energy of the W particles is 5–6 times higher than that of SiC particles of the same size. This is due to their difference in density (W 19.3 g/cm^3, SiC 3.22 g/cm^3). Additionally, a mutual deformation of W and Ni particles occurs. Besides this, there are considerably less pores at the particle–matrix interface. This is clear evidence for the change

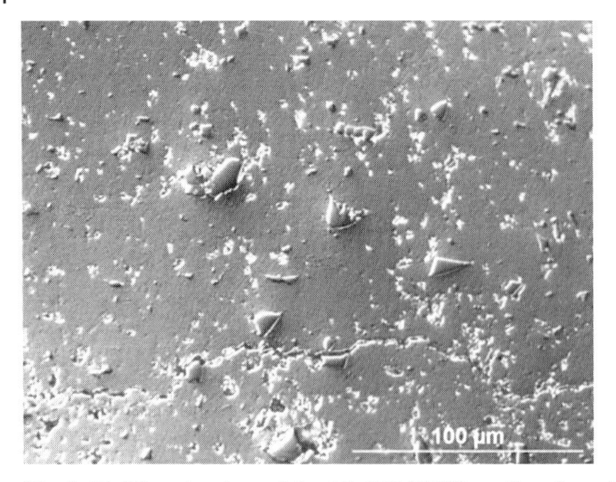

Fig. 9.13 Microstructure of the Ni–SiC (25%) coating deposited at an air temperature of 450°C.

in particle deformation and densification mechanisms during GDS. The image in Fig. 9.14(a) shows a near-full density area in the coating. Here the tungsten particles are surrounded by the dense Ni matrix. The densification process is fully realized when the tungsten particles are uniformly distributed in the nickel matrix.

9.3.2.2 **Microhardness and Microscratching**

It is difficult to observe the interparticle boundaries of the Ni–SiC and Ni–W coatings due to severe particle deformation (Fig. 9.13 and 9.14). For this reason, examination of the microscratch behavior and coating–substrate interface morphology by SEM and optical microscopy was performed on the layers of soft Ti (HV 97–100) and hard Inconel (HV 450) particles deposited on Al substrate (HV 80) by GDS. The images are shown in Figs. 9.15, 9.16 and 9.17. The results reveal that the hard Inconel particles act as penetrators, causing the intensive flow of the soft Al substrate. The depth of penetration into the Al substrate is significant. This validates the assumption that considerable localization of the deformation in the Al substrate takes place at the interface due to the adiabatic shear band formation process. The Al substrate is subsequently deformed by secondary particles, resulting in the pores around the penetrating particles in the Al substrate being closed. This process depends on the kinetic energy of the particles.

The lamellar morphology of the Al substrate is clearly seen from the images shown in Fig. 9.16, as well as in the results of the microhardness measurements. The microhardness of the Al substrate near the interface is about 1.18 GPa (HV = 121, Fig. 9.15(b)) – an increase in micohardness by 1.5 times. Ti particles are deformed and hardened during GDS as well (Figs. 9.15 and

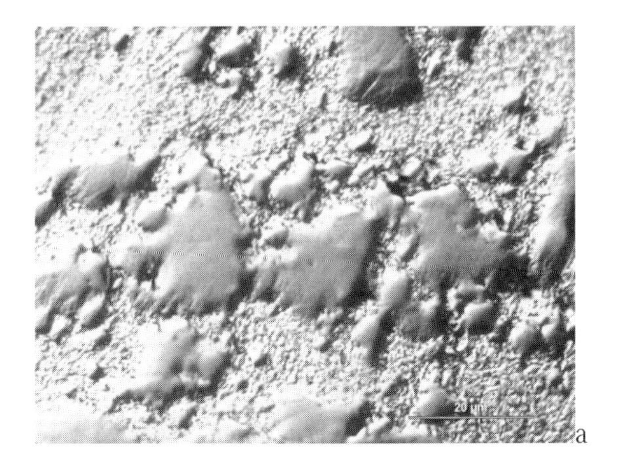

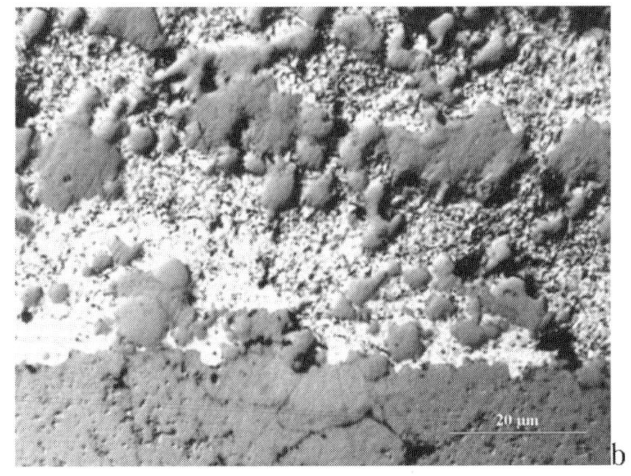

Fig. 9.14 Microstructures of the Ni–W (25%) coating deposited at an air temperature of 450°C. (a) Coating center; (b) coating–substrate interface.

9.17), with their microhardness measuring 2.15–3.1 GPa. The microhardness of Ti particles increases by 2–3 times due to strain hardening. The localization of deformation within the Ti particles results in different hardness values (Fig. 9.16). These data reveal that the impact deformation of the particles and substrate has a considerable effect on the formation of the GDS coating structure.

Instrumented scratch testing is known to be a useful tool in characterizing the abrasion resistance of a material (Williams 1996). Whereas, hardness testing has been intensely studied and modeled over the last decades, a complete deterministic model for scratching does not exist. This is a result of the fact that scratch testing is most often used as a quality control technique for comparing the surface mechanical performance between different materials (Williams 1996). It is very important in our case, however, to characterize

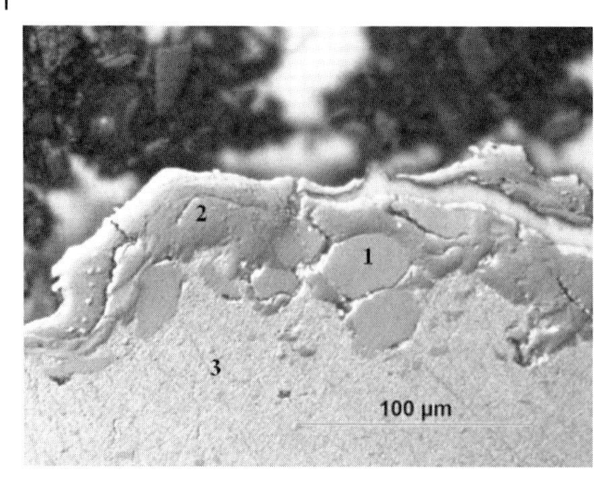

Fig. 9.15 Microstructure of the deposited layer on an Al substrate (GDS air temperature 450°C). (1) Inconel particles; (2) Ti particles; (3) Al substrate.

the mechanical performance of each of the constituents in the GDS composite coating. For this reason further model development is necessary.

The characteristic parameter evaluated in a scratch test is the scratch hardness H_s, which is calculated by the relationship $H_s = P/A$, where P is the normal load and A is the projected load bearing area $A = w^2/4$ for Vickers indenter, where w is the width of the residual scratch. The width of the residual scratch depends on both the material and interface properties. A comparison of microhardness and scratch hardness parameters is shown in Fig. 9.18. The values of microhardness and scratch hardness don't differ considerably for the soft Al substrate, contrary to that of Ti and Inconel particles. In fact, these data show that these particles act to increase the wear resistance of such composite coatings. Further, the wear resistance may be controlled by the type and content of deposited particles through the determination of particle scratch hardness.

Optical microscopy images reveal the presence of a lamellar morphology and a strong interparticle bonding in the composite coating (Figs. 9.16 and 3.8). An analysis of the experimental data further demonstrates the retention of nanoscale structure as well as localized grain refinement, which is thought to result from adiabatic shear bands. The primary conclusions that may be drawn from this analysis are:

1. The densification process is fully realized with Ni–W powder mixtures when the tungsten particles are uniformly distributed in the nickel matrix.
2. Optical microscopy observations reveal the presence of a lamellar morphology and a strong interparticle bonding in the composite coating.

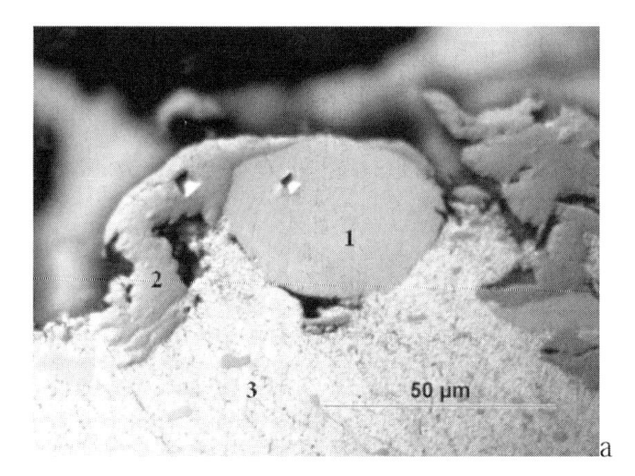

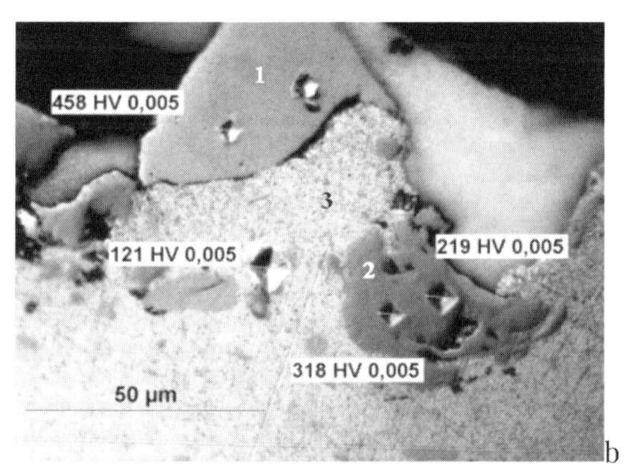

Fig. 9.16 Microstructure and microhardness of the deposited particles on an Al substrate (GDS air temperature 450°C). (1) Inconel particles; (2) Ti particles; (3) Al substrate.

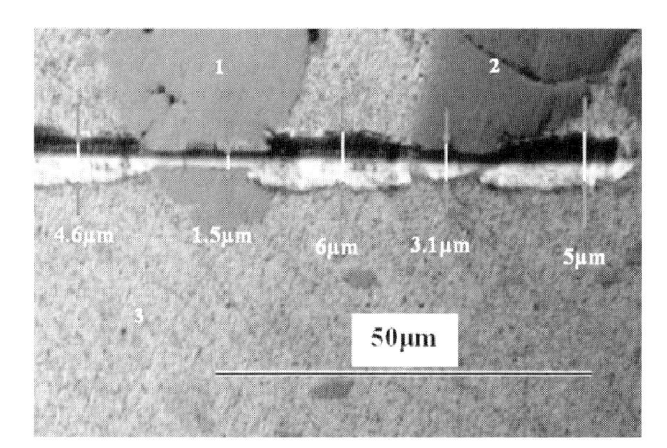

Fig. 9.17 Microstructure and microscratch track of the particles on Al substrate (GDS air temperature 450°C, normal indenter load 0.01 N). (1) Inconel particles; (2) Ti particles; (3) Al substrate. The red arrows indicate the track width.

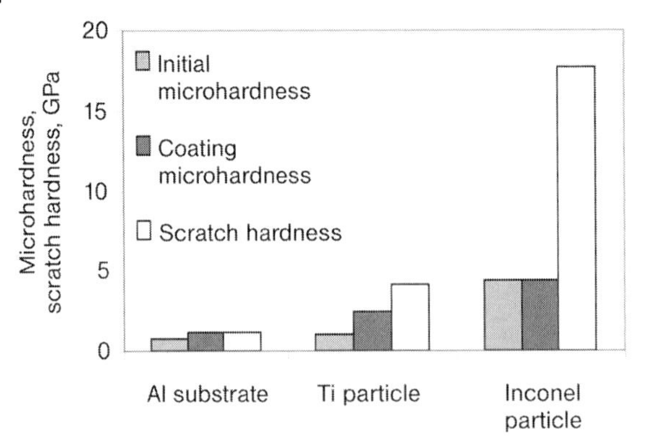

Fig. 9.18 The effect of GDS on the microhardness and scratch hardness of the Al substrate, Inconel, and Ti particles.

9.4 Corrosion Protection Through GDS Deposition

9.4.1 General Remarks

Corrosion is another difficulty that must be overcome within industrial and manufacturing environments. Many parts that have been damaged by corrosion or corrosive wear are currently being repaired and restored to their original dimensions using thermal spraying techniques (Jones 1992). By rapidly depositing a coating with a thickness of several 100 μm or more to the surface, thermal spraying can provide the necessary volume of material with sufficient adhesion strength to build up the damaged surface. Applications include chemical tank liner, anticorrosion/erosion coatings for pipelines and corrosion resistant coatings for various components in harsh environments. These thermal spraying techniques, including electric arc spraying and combustion flame spraying are largely employed when thick coatings (¿0.5 mm) are needed. Thermal spraying is also used to upgrade component materials by applying a high-grade corrosion-resistant coating – including Ni- or Co-based alloys on steel or stainless steel substrates – to improve their anticorrosion properties (Niranatlumpong and Koiprasert 2006).

Due to their good tribological and corrosion–oxidation properties in aggressive environments, thermally sprayed Cr_3C_2-NiCr coatings are widely used in aircraft manufacturing, the automotive industry, paper manufacturing machinery, and other engineering applications. In fact, they are often employed as an alternative to hard Chromium plating in applications where both wear and corrosion resistance are required (Suegama et al. 2006). The application of GDS techniques for corrosion protection has been developed by McCune et al. (1996, 2000), Karthikeyan et al. (2004), and others. Karthikeyan et al. (2004) state that cold-sprayed Al coatings can be applied to protect metal surfaces from mild atmospheric degradation by means of a very thin and impervious oxide layer being formed on an aluminum surface.

Because GDS can produce very thick coatings in a short period of time, it is widely used in component refurbishment, where substantial surface build-up is necessary. The inherent porosity of LPGDS coatings, however, inhibits their use as a corrosion prevention method. In fact, coatings produced via LPGDS technique often contain moderate to high porosity, ranging from 3 to 18%. These pores are commonly interconnected. Thus, molecular transport of corrosive agents through these pores is likely. This can accelerate the corrosion failure of the coating via larger surface contact and/or crevice mechanisms. In severe cases, the corrosive agents may gain access to the coating–substrate interface and cause the corrosion of underlying components, leading to part failure.

In order to fully evaluate the degree of protection offered by various coating systems it is first necessary to differentiate the types of protection that these coatings are able to provide. The main type of protection that GDS coatings afford is a barrier to water and ion ingress. This type of protection works by effectively sealing off the substrate from the corrosive environment, and can be examined by analyzing intact areas of coating samples. However, as previously shown, barrier protection becomes compromised in areas of the coating containing defects or pores. Protection in defect areas can be provided by sacrificial coatings which corrode preferentially with respect to the substrate. As an example, Zn is more active than steel in most environments and therefore is commonly used as a sacrificial coating, providing cathodic protection. This form of protection can be analyzed by examining intentional defects in coatings (such as scribes) and running electrochemical tests on these areas. Both barrier and sacrificial protections are important when determining what type of coating to use in a certain application.

9.4.2 Examination of Al–Zn-based Sacrificial Coatings

Aluminum, being a low-density material, acts as a good spray material as it easily achieves the spray velocities that are required for deposition (Tokarev et al. 1996). It is also ductile and therefore can easily undergo sufficient plastic deformation when impacting the substrate, in turn bringing about effective bonding. Moreover, if a deformed particle leaves a pore after deposition, further impact and plastic deformation of subsequent aluminum particles tends to close the pores and voids of the coating (Smith et al. 1999). It is for these reasons that the repair of aluminum structures using cold spray is highly attractive for reclamation purposes.

To test the level of protection afforded by GDS coatings, $Al–Al_2O_3$, $Zn–Al–Al_2O_3$ and Zn–Al–Ti powder mixtures (particle size of 40–50 μm) were sprayed by LPGDS on two substrates, one Al and one mild steel. The main LPGDS deposition parameters were as follows: air pressure, 5–6 bars; air consumption, 0.5–1 m^3/min; particle velocity, 400–500 m/s; particle feeding rate, 0.2–0.6 g/s.

Electrochemical polarization tests of cold-sprayed $Al–Al_2O_3$ coatings on substrates of the same composition have been carried out at 0.5 N H_2SO_4. The exposed area in electrolyte for the samples had dimensions of 30 × 30 × 5 mm. All the experiments were carried out after polishing the samples with 600-grit silica abrasive. The free corrosion potential (E_{cor}) of the coated sample was measured to define the range of experimental potentials in the electrochemical experiments.

Salt spray (fog) testing is the most popular form of testing for protective coatings (ASTM G5, ASTM G59). These tests have been used for more than 90 years to determine the degree of protection afforded by both inorganic and organic coatings on a metallic substrate. The most widely used salt spray (fog) tests are described below. The neutral salt spray (fog) test (ASTM B 117) is perhaps the most commonly used salt spray test in existence for testing inorganic and organic coatings, in particular when determining material or product specifications. The duration of this test can range from 8 to over 3000 h, depending on the product. A 5% sodium chloride solution containing not more than 200 ppm (parts per million) total solids and with a pH range of 6.5–7.2 is used. The temperature of the salt spray chamber was controlled to maintain 35°C within the exposure zone of the closed chamber.

Some of the results of these corrosion tests are shown in Figs. 9.19–9.21, where the main conclusions are as follows:

1. The sacrificial properties of GDS Zn–Al-based coatings are sufficient to achieve effective corrosion protection of steel substrates with these coatings.
2. The coating potential of $Al–Al_2O_3$ does not differ from that of the Al substrate.
3. The effectiveness of Zn–Al-based LPGDS coatings on the engine blocks are clearly demonstrated in real engine block corrosion tests (Fig. 9.21).

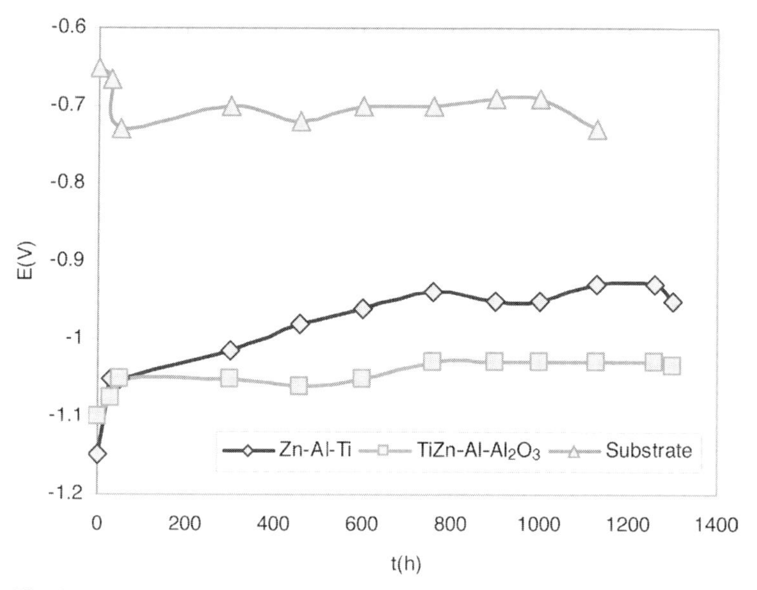

Fig. 9.19 Relationship between potential and immersion time for samples in 5%NaCl solution.

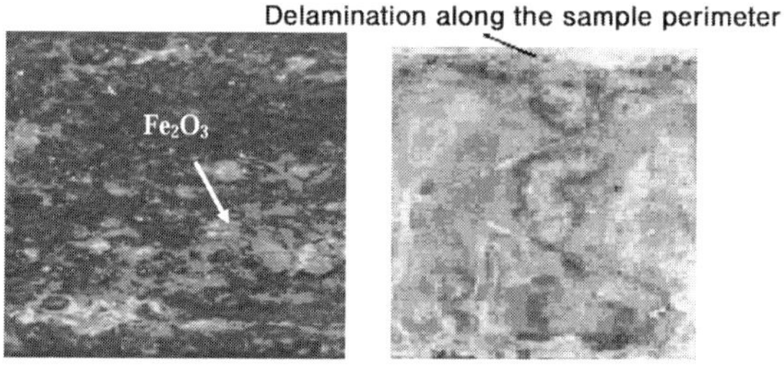

Delamination along the sample perimeter

Fe_2O_3

Fig. 9.20 Steel substrate after 1000 h exposure in 5%NaCl solution. 50%Zn–40%Al–10%Al_2O_3 coating after 1000 h exposure in 5%NaCl solution.

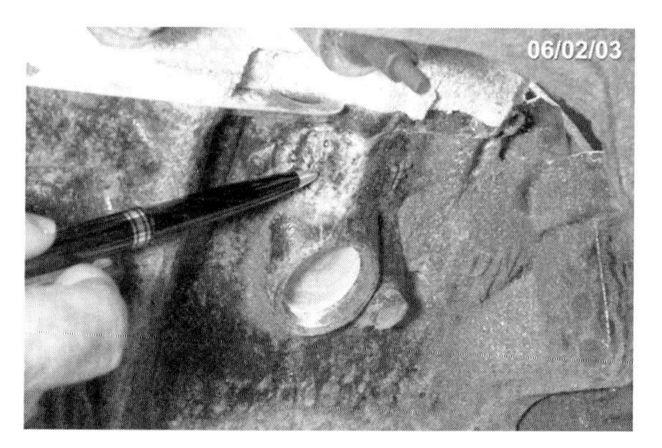

Fig. 9.21 Results of an engine block corrosion test.

9.5 GDS Processing of Smart Components

9.5.1 General Remarks

During recent decades, the development of new metal powders and processes has led to a considerable expansion of powder metallurgy (PM) applications. High-performance sintered components demand light powder materials with improved mechanical properties at reduced costs. The automotive industry, as the largest consumer of PM parts, is the driving force behind the investigation of materials for lightweight components. In recent years, some PM aluminum alloys have been optimized and powder mixtures equivalent to the wrought alloy AA2014 have found their way into mass production (Arnhold et al. 2004).

Large-scale development of lightweight components is still a challenge for the PM industry. Unfortunately, classical PM technologies such as compaction-sintering do not result in lightweight alloys with sufficiently high densities (Roseto et al. 1991). For this reason, the development of new high-density powder technologies for Al alloys is of great importance. GDS permits complex thin-walled shapes of various powder materials and composites. Further, optional postspraying processes such as sintering (Calla et al. 2005), sizing and minor machining may be readily applied. Using the low-pressure radial injection GDS method, it is possible to form various novel thin wall components. One particular benefit of spray technology in this application is the weight reduction that is accompanied by the use of aluminum instead of steel. Also, in the near future spray powder technologies may more readily enable large scale production of components.

One method to decrease the weight of components is simply to replace commonly used PM steel with PM aluminum. It is reasonable to develop a new GDS spraying technology for the net shape forming of PM Al composites. In turn, it is beneficial to compare this with PM aluminum alloys, which combine adapted physical properties with the new near net shape technology. Recently, the University of Windsor (Windsor, Ontario, Canada) developed new low-pressure GDS technology of aluminum compounds for thin wall bushing applications, as well as a new GDS-sintering sizing processing route for aluminum composites (Weinert et al. 2006). It is, therefore, the aim of this section to compare the structure and properties of Al-based composites made by GDS with the same composites made by conventional PM technology.

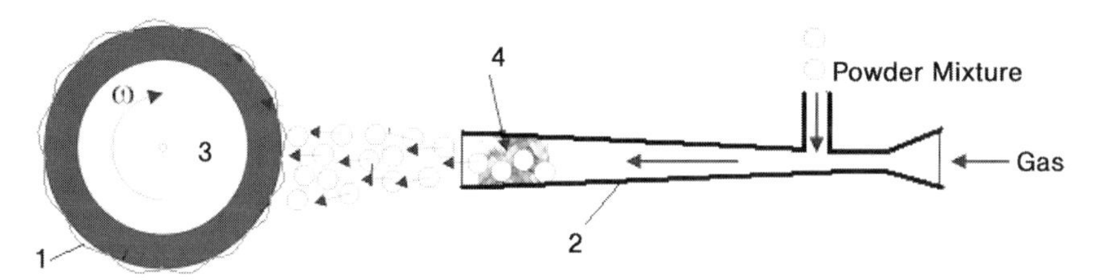

Fig. 9.22 GDS net shape forming of cylindrical components. (1) Sample of bushing; (2) DYMET gun; (3) mandrel; (4) accelerated powder mixture.

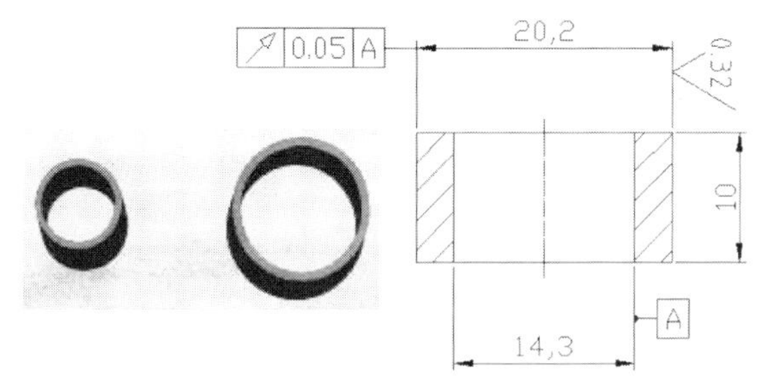

Fig. 9.23 Images and drawing of an experimental bushing.

9.5.2 Technology Description

The experiments were conducted using Al and Al_2O_3 powders, the characteristics of which are given in Table 9.5. In this case, fine Al_2O_3 powders were used as reinforcement particles to strengthen the Al matrix. In particular, an Al–Al_2O_3 (15 wt.%) powder mixture was used.

Tab. 9.5 Characteristics of Al and Al_2O_3 powder supplied by Atlantic Equipment Engineers Co.

Name of the powder	Composition			Particle size (μm)	Apparent density (g/cm^3)
	Al	Mo	Fe		
Al	base	0.01	0.03	$\leq 45(30\%)$ to $\leq 150(70\%)$	1.25
Al_2O_3	case	–	0.03	$\leq 5(20\%)$ to $\leq 10(80\%)$	1.65

The gun was installed on a gun holder to scan the air-powder jet over the cylinder surface of the sample (Fig. 9.22). Images and a drawing of the bushing samples are shown in Fig. 9.23.

The powder compaction and sizing experiments were carried out using special tooling, which was installed on the test machine. The density of the compact was measured by Archimedes' method after the compaction and forging–repressing. The operations were performed by means of an INSTRON test machine with load capacity of 200 kN. A schematic of the forging tool set up on

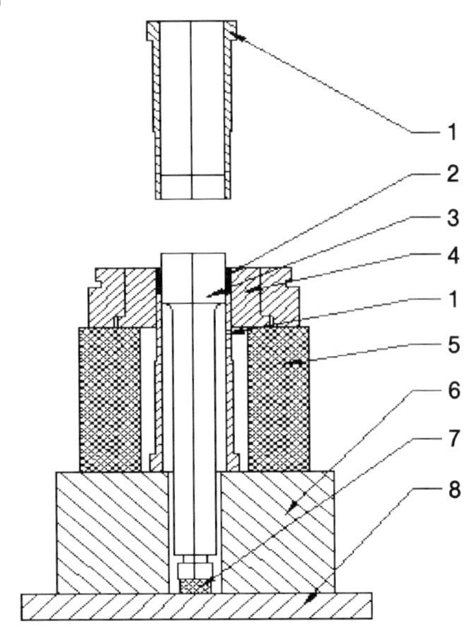

Fig. 9.24 The scheme of the sizing tool set up on an INSTRON press. (1) Punch; (2) bushing sample; (3) rod; (4) die with radial load sell; (5) matrix plate of the die; (6) distance sleeve; (7) load sell of the rod; (8) base plate.

the INSTRON press is shown in Fig. 9.24. The load–displacement diagrams were recorded via computer and corrected to take into account the elastic deformation of tooling. The normal load that was applied to the upper and lower punches, as well as the radial stresses, was recorded during the forging operation. Powder rings and compression samples with sizes of $\oslash 10$ mm $\times$ 10 mm, composed of Al–Al_2O_3, were used for the sizing operation.

Samples of Al–15%Al_2O_3 composite were made by GDS and usual PM compaction technology. After sintering at a temperature of 600°C, the PM samples were forged to a density of about 95%, while the GDS samples were sized with small strains (10–12%). Each sample was also homogenized at 500°C for 1 h. In the case of PM compaction, a Zn stearate binder was added to the composition of the powder mixture (Table 9.6).

Tab. 9.6 Composition of powder mixture for PM compaction.

Al (wt.%)	Cu (wt.%)	Al_2O_3 (wt.%)	StZn (wt.%)
Base	2	15	0.8

Compression tests were performed using the INSTRON machine. An example of the load–displacement diagram and an image of the sample is shown in Fig. 9.25. The yield and fracture strengths were determined using compression diagrams (Fig. 9.25) and a microhardness tester MP-100TC was used

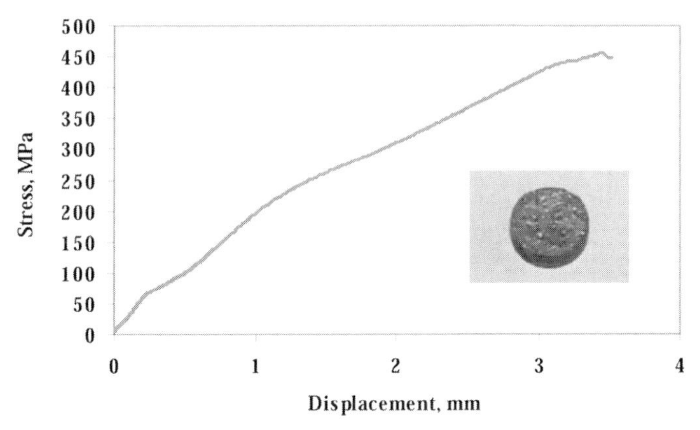

Fig. 9.25 Compression load–displacement diagram and an image of the sample.

for microhardness measurements. The sample structure was examined using a light microscope (Nicon EclipseL 150) after polishing and etching with Miller's reagent for 10 s

Conventional processes for PM aluminum components of ⌀10mm × 10 mm size consist of the following steps:

1. compaction to a green density of approximately 2.5–2.6 g/cm^3;
2. dewaxing;
3. sintering;
4. homogenization;
5. sizing;
6. final heat treatment to adjust specific material properties.

From a chemical point of view, most PM aluminum alloys are based on common cast or wrought alloys of 2000, 6000, and 7000 series (Arnhold et al. 2004). The alloying elements of Cu, Mg, Si, and Zn are admixed as pure elements or in the forms of masteralloys (Arnhold et al. 2004). The masteralloy is an alloy containing at least some aluminum and one or more added elements for use in making alloying additions to molten aluminum. As it turns out, complete alloying typically results in unsatisfactory compressibility and low sinter activity. Pure aluminum powder and mixtures of aluminum and copper show better compressibility than iron-based mixtures (Lall and Heath 2000, Simchi and Veltl 2003). The compressibility is reduced by admixed brittle and hard masteralloys, especially for the compaction of geometries with a high ratio between sliding surfaces and pressing faces. This problem can be solved, however, by either introducing increased lubrication or by warm compaction of the mixture (Simchi and Veltl 2003). In order to achieve the desired mechanical properties of the sintered parts, however, it is necessary to remove the lubricant before sintering. Also, the dewaxing temperature is recommended to be below 420°C to avoid liquid phase formation during dewaxing (Arnhold et al. 2004). The dewaxing is normally complete in about 20 min in a nitrogen atmosphere. Due to the natural oxide skin on the aluminum particles, sintering can only be achieved by the presence of liquid phases (Savitskii 1993, German 1985). These eutectic phases built by aluminum and the alloying elements penetrate the cracks in the oxide layers. The alloying elements of the liquid phase diffuse into the aluminum particles, forming a homogeneous alloy. Similar to aluminum cast and wrought alloys, PM alloys exhibit their excellent mechanical behavior only after heat treatment. This treatment begins with homogenization at an increased temperature, followed by fast cooling (quenching) and a final treatment at lower temperatures (age hardening) (Simchi and Veltl 2003, German 1985).

Various small inclusions from the oversaturated solution precipitate in the last step of the heat treatment, causing increased strength and material hardness. Dimensional changes during sintering are caused by the presence of the liquid phase. Irregularities in mold filling (density gradient), temperature gradients during sintering and changes in atmosphere (dew point) can result in deformation of the PM aluminum components. For most applications it is therefore necessary to size the parts appropriately to acquire the required tolerances. Another aspect is that during sizing, most aluminum components obtain a higher density and mechanical performance.

The application of GDS as the main compaction technology allows high-density components to be generated in a single step (Fig. 9.22). In fact, gas dynamic spraying yields near full density deposited layers (Papyrin 2001, Assadi et al. 2003). It is obvious that complete sintering is not realized during GDS, however, these processes occur partially. Therefore, a combination of GDS forming operation with sintering and sizing allows one to overcome the above described disadvantages of conventional PM processes for Al components.

9.5.3
Results and Discussion

The results of mechanical tests and porosity measurements of samples made by the two processing routes of GDS forming and PM compaction are shown in Table 9.7. It can be seen that the new GDS forming process leads to microstructures with very low-residual porosity ($>$ 97% of theoretical density), as is usual for powder-forged Al alloys. The density of GDS formed samples is higher than those of PM compacted samples by 1–2%. The yield compression strength data reveal the positive effect of high-particle velocity on the mechanical properties of Al–15%Al_2O_3 composites.

Although the properties of wrought Al-based composites are very nearly achieved, this data reveals lower ductility, which is not satisfactory for appli-

Tab. 9.7 Results of compression tests and density measurement.

Technology	Porosity (%)	Yield strength (MPa)	Fracture strength (MPa)	Fracture strain (%)
	2.3	220.3	253.7	32
GDS – sintering–sizing–	3.2	200.8	201.50	16
homogenization	2.1	238.5	269.9	12
	2.8	245.7	287.7	35
Compaction – (I)	3.7	182.3	220.4	19
sintering–forging; (II)	4.1	191.2	203.3	16
sintering–sizing–	2.9	175.8	211.3	22
homogenization	2.8	183.7	209.55	12

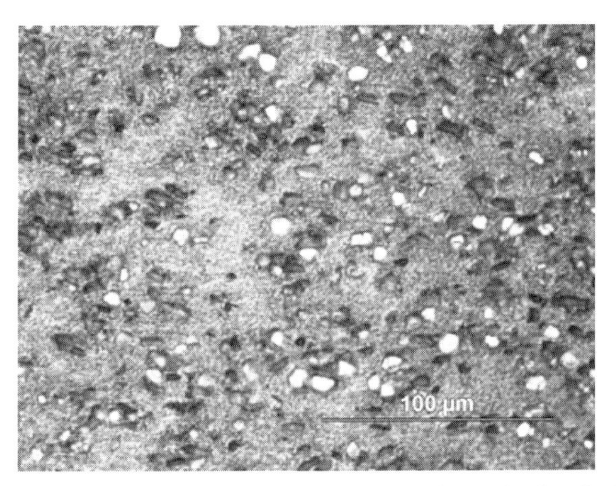

Fig. 9.26 Microstructure of PM compacted sample after final operation. Compaction pressure, 400 MPa; forging pressure, 800 MPa. Some pores may be observed (dark fields).

cation on structural components in heavy loaded applications. The mechanical properties formed GDS formed specimens are 20–30% higher than created using conventional processes.

Figures 9.26 and 9.27 show the respective microstructures of GDS formed and conventional PM compacted samples. The primary difference between these images can be seen in the anisotropy of the reinforcement particle distribution in the case of GDS formed specimens. An analysis of GDS formed composite microstructures after etching (Fig. 9.27) confirms that the anisotropy of the particle distribution is a result of localized deformation during the spraying process. A significant amount of highly deformed Al particles are also clearly seen (Fig. 9.27(a)). It should also be noted that the position of the Al_2O_3 particles is strongly determined by the flow direction of the Al matrix.

In the case that the Al matrix flow is turbulent, a random distribution of the reinforcement particles is achieved (Fig. 9.27(b)). However, the structure

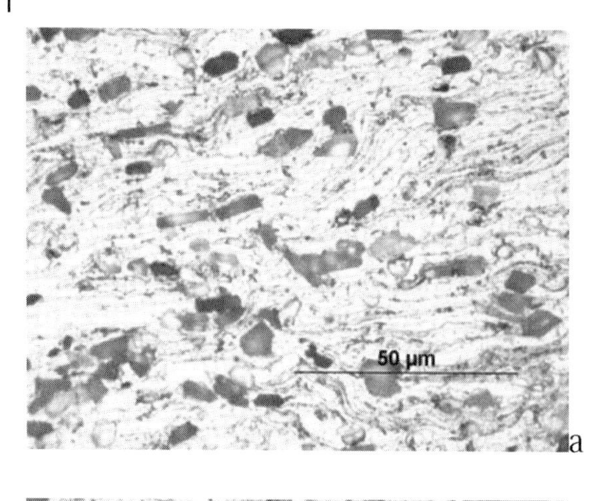

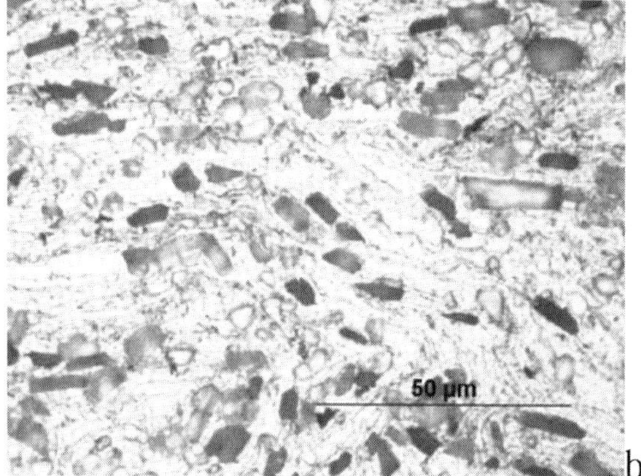

Fig. 9.27 Microstructure of GDS formed sample after sintering and etching by Miller's reagent. (a) Laminar flow pattern; (b) turbulent flow pattern.

and properties of PM aluminum-based composites may be improved through GDS processing because such an intensive flow of the soft matrix results in the rupture of the surface aluminum oxide films, increasing the number of direct metallic contact areas after spraying and subsequent sintering. An AISI 316 stainless steel mandrel substrate was used to make bushings with a wall thickness of 1.5–3 mm (Fig. 9.23). The hardness distribution in the bushing wall ranges from 90 to 110 $HV_{0.1}$, as shown in Fig. 9.28.

The introduction of a new GDS forming process for net shape PM aluminum components with improved mechanical behavior provides an opportunity for further development. Further investigations will concentrate on processing, microstructure, and the mechanical behavior of wear resistant PM aluminum alloys and Al matrix composites.

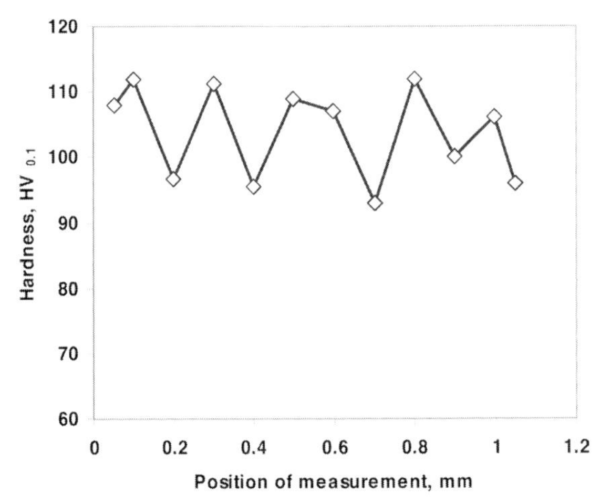

Fig. 9.28 Hardness distribution on the cross section of a GDS formed thin wall bushing made of an $Al–Al_2O_3$ powder mixture.

Thus, the introduction of a GDS step into the process in the net shape forming of high-density Al-based composite components results in an overall improvement in the material's performance. This superior behavior is primarily the result of intensive particle deformation during GDS forming. Also, the anisotropy of the particle distribution is a result of localized deformation during the spraying process.

9.6 Concluding Remarks

The examples of LPGDS technology mentioned above demonstrate the flexibility of this process within industrial environments. Some powder mixture recipes were developed to comply with the technical requirements of real components.

The friction and wear behaviors of Cu-based W-reinforced composite coatings and Ni-based TiC-reinforced composite coatings deposited on cast iron and steel substrates were shown to be similar to those of cast iron and steel substrate materials.

The dry sliding characteristics were shown to be dependent on the applied load and structure of the composite coatings. For the applied loads, wear occurred by a combination of adhesion, delamination, and ploughing with a high (0.65) friction coefficient. The use of GDS to apply composite coatings compatible with cast iron and steel was found beneficial for applications in cast iron die component repair and corrosion protection. The particular friction and wear mechanism is related to the characteristics of the transfer layer which is being formed at the interface during sliding.

Bibliography

1 Abramovich G.N., (1963), *The Theory of Turbulent Jets*, MIT Press, Cambridge, MA.

2 Alkhimov A.P., Gulidov A.I., Kosarev V.F., Nestrovich N.I., (2000), Specific features of microparticle deformation upon impact on a rigid barrier, *J. Appl. Mech. Tech. Physics*, 41(1), pp. 204– 209.

3 Alkhimov A.P., Kosarev V.F., Klinkow S.V., (2001), The features of cold spray nozzle design, *J. Therm. Spray Technol.*, 10, pp. 375–381.

4 Alkimov A.P., Kosarev V.E., Papyrin A.N., (1990), A method of cold gas-dynamic deposition, *Dokl. Akad. Nauk SSSR*, 318, pp. 1062–1065.

5 Alkhimov A.P., Papyrin A.N., Kosarev V.F., Nesterovich N.I., Shushpanov M.M., (1994), Gas-dynamic spraying method for applying a coating, US Patent 5,302,414.

6 Amateau M.F., Eden T.J., (2000), High-velocity particle consolidation technology, *iMAST Quarterly*, 2, pp. 3–6.

7 Anderson T.L., (1995), *Fracture Mechanics: Fundamentals and Applications*, CRC Press, Boca Raton, FL.

8 Araujo P., Chicot D., Staia M., Lesage J., (2005), Residual stresses and adhesion of thermal spray coatings, *Surf. Eng.*, 21, pp. 35–39.

9 Arnhold V., Balzer H., DeCristofaro N., Kruzhanov V., Laux M., Lindenau R., (2004), Development of powder metallurgical aluminium alloys and applied processes, *PM 2004*, (CD Proceeding), Viena, Austria, Vol. 4, pp. 78–83.

10 Arzt E., (1998), Size effects in materials due to microstructural and dimensional constraints: a comparative review, Overview No. 130, *Acta Mater.*, 46(16), pp. 5611–5626.

11 Assadi H., Gärtner F., Stoltenhoff T., Kreye H., (2003), Bonding mechanism in cold gas spraying, *Acta Mater.*, 51, pp. 4379–4394.

12 ASTM Designation E8-01, *Standard Test Method for Tension Testing of Metallic Materials*, Philadelphia, 2001.

13 Atkinson, J. H., *An Introduction to the Mechanics of Soils and Foundations: Through Critical State Soil Mechanics*, McGraw-Hill, London, New York, 1993, p. 337.

14 Bai Y., (1982), Thermo-plastic instability in simple shear, *J. Mech. Phys. Solids*, 30, pp. 195–201.

15 Barbezat, G. (2006), Application of thermal spraying in the automobile industry, *Surf. Coat. Technol.*, 201, pp. 2028–2031.

16 Bateni M., Szpunar J.A., Wang X., Li D.Y., (2006), Wear and corrosion wear of medium carbon steel and 304 stainless steel, *Wear*, 260, pp. 116–122.

17 Beneteau M., Birtch W., Villafuerte J., Paille J., Petrocik M., Maev R.Gr., Strumban E., Leshchynsky V., (2006), Gas dynamic spray composite M. coatings for iron and steel castings, *International Thermal Spray Conference Proceedings*, Seattle, USA, (CD Proceeding).

18 Benson D.J., Nesterenko V.F., Jonsdottir F., Meyers M.A., (1997), Quasistatic and dynamic regimes of granular material deformation under impulse loading, *J. Mech. Phys. Solids*, 45(11/12), pp. 1995–1999.

19 Bhagat R.B., Amateau M.F., Papyrin A.N., Conway J.C.J., Stutzman B.J., Jones B., (1997), in: C.C. Berndt (Ed.), *Thermal Spray: A United Forum for Scientific and Technological Advances*, ASM International, Materials Park, OH, pp. 361–367.

20 Bialucki P., Kozerski S., (2006), Study of adhesion of different plasma-sprayed coatings to aluminium, *Surf. Coat. Technol.*, 201, pp. 2061–2064.

21 Billig F.S., (1967), Shock-wave shapes around spherical and cylindrical-nosed bodies, *J. Spacecr.*, 4(5), pp. 822–823.

22 Borchers C., Gärtner F., Stoltenhoff T., Kreye H., (2003), Microstructural and macroscopic properties of cold sprayed copper T. coatings, *J. Appl. Phys.*, 12 93(2), pp. 10064–10070.

23 Borchers C., Gärtner F., Stoltenhoff T., and Kreye H., (2004), Microstructural bonding features of cold sprayed face centered cubic metals, *J. Appl. Phys.*, 96(8), pp. 4288–4292.

24 Borchers C., Gärtner F., Stoltenhoff T., Kreye H., (2005), Formation of persistent dislocation loops by ultra-high strain-rate deformation during cold spraying, *Acta Mater.*, 53, pp. 2991–3000.

25 Calla E., McCartney D.G., Shipway P.H., (2005), Effect of heat treatment on the structure and properties of cold sprayed copper, *ITSCE 2005*, Proceeding (CD) Basel, Switzerland.

26 Carlson D., Hoglund R., (1964), Particle drag and heat transfer in rocket nozzles, *AIAA J.*, 2(11), pp. 1980–1984.

27 Carroll M.M., Holt A.C., (1972), Static and dynamic porecollapse relations for ductile porous materials, *Journ. Appl. Phys.*, 43(4), pp. 1626–1636.

28 Champagne V., Helfritch D., Leyman P., Lempicki R., Grendahl S., (2005), The effects of gas and metal characteristics on sprayed metal coatings, *Model. Simul. Mater. Sci. Eng.*, 13, pp. 1119–1128.

29 Chen Li, Batra R.C., (2000), Microstructural effects on shear instability and shear band spacing,*Theor. Appl. Fract. Mech.*, 34, pp. 155–166.

30 Choi D.S., Lee S.H., Shin B.S., Whang K.H., Song Y.A., Park S.H., Jee H.S., (2001), Development of a direct metal freeform fabrication technique using CO_2 laser welding and milling technology, *J. Mat. Process. Technol.*, 113(1–3), pp. 273–279.

31 Clobes J.K,. Green D.J, (2002), Validation of single-edge V-notch diametrical compression fracture toughness test foe porous alumina, *J. Mater. Sci.*, 37, (2002), pp. 2427–2434.

32 Conzone S.D., Butt D.P., Bartlett A.H., (1997), Joining $MoSi_2$ to 316 stainless steel, *Journ. Mat. Sci.*, 32(13), pp. 3369–3374.

33 Conteras A., Albiter A., Bedolla E., Perez R, (2004), Processing and characterization of Al–Cu and Al–Mg base composites reinforced with TiC, *Adv. Eng. Mater.*, 6, p. 9.

34 Crossland B., (1982), *Explosive Welding of Metals and its Application*, Claredon Press, Oxford.

35 Cullison, A. (2001), Stress relief basics, *Welding J.*, 80(9), p. 49.

36 Dalley H.S., (1970), Powder feed device for flame spray guns, US Patent 3,501,097.

37 Daridon L., Oussouaddi O., Ahzi S., (2004), Influence of the material constitutive models on the adiabatic shear band spacing: MTS, power law and Johnson-Cook models.

38 Date H., Kobayakawa S., Naka M., (1999), Microstructure and bonding strength of impact-welded aluminium–stainless steel joints, *J. Mater. Process. Technol.*, 85 (1–3)166; *Int. J. Solids Struc.*, 41, pp. 3109–3124.

39 DiLellio J.A., Olmstead W.E., (1997), Temporal evolution of shear band thickness, *J. Mech. Phys. Solids*, 45, pp. 345–359.

40 DiLellio J.A., Olmstead W.E., (2003), Numerical solution of shear localization in Jonson–Cook materials, *Mech. Mater.*, 35, pp. 571–580.

41 Djordjevic B., Maev R., (2006), SIMAT Application for aerospace corrosion protection and structural repair, *International Thermal Spray Conference Proceedings*, Seattle, USA, (CD Proceeding).

42 Dodd B., Bai Y., Width of adiabatic shear bands formed under combined stresses, *Mater. Sci. Technol.*, 1989, 5(6), pp. 557–559.

43 Dykhuizen R.C., Neiser R.A., (2003), Optimizing the cold spray process, in: *Proceedings of the International Thermal Spray Conference*, ASM International, Materials Park, OH, USA, Orlando, 5–8 May, pp. 19–26.

44 Dykhuizen R.C., Smith M.F., (1989), Investigations into the plasma spray process, *Surf. Coat. Technol*, 37(4), pp. 349–358.

45 Dykhuizen R.C., Smith M.F., (1998), Gas dynamic spray principles of cold spray, *J. Therm. Spray Technol.*, 7(2), pp. 205–212.

46 Dykhuizen R.C., Smith M.F., Gilmore D.L., Nelser R.A., Jiang X., Sampath S., (1999), Impact of high velocity cold spray particles, *J. Therm. Spray Technol.*, 8(4), 559–564

47 El-Sobky H., (1983), Mechanics of explosive welding, in: T.Z. Blazynsky (Ed.), *Explosive Welding, Forming and Compaction*, Applied Science Publishers, London

48 Eskin D., (2005), Modeling dilute gas-particle flows in horizontal channels with different wall roughness, *Chem. Eng. Sci.*, 60, pp. 655–663.

49 Eskin D., Kalman H., (2002), Engineering model of friction of gas–solids flow in a jet mill nozzle, *Chem. Eng. Technol.*, 25, pp. 57–64.

50 Eskin D., Voropaev S., (2001), Engineering estimations of opposed jet milling efficiency, *Miner. Eng.*, 14, pp. 1161–1175.

51 Eskin D., Voropaev S., (2004), An engineering model of particulate friction in accelerating nozzles, *Powder Technol.*, 145, pp. 203–212.

52 Eskin D., Voropaev S, Dorokhov I., (2003), Effect of particle size distribution on wall friction in high-velocity gas-dynamic apparatuses, *Theor. Found. Chem. Eng.*, 37(2), pp. 122–130.

53 Eskin D., Voropaev S., Vasilkov O., (1999), Simulation of jet milling, *Powder Technol.*, 105, pp. 257–265.

54 Evans J.F., Kirkwood D.H, Beech J., (1991), in: M. Rappaz, M.R. Ozgu, K.W. Mahin (Eds.), *Modeling of Casting, Welding and Advanced Solidification Processes V*, TMS, Warrendale, PA, p. 533.

55 Fabel A., (1976), Powder feed device for flame spray guns, US Patent 3,976,33.

56 Flemming R.P., Olmstead W.E., Davis S.H., Shear localization with an Arrhenius flow law, *J. Appl. Math.*, 2000, 60(6), pp. 1867–1886.

57 FLUENT 1996, 4.4.4 User Guide, Fluent Inc., 1996.

58 Fuchs N.A., (1997), *The Mechanics of Aerosols*, Pergamon Press, New York, 1964.

59 Fukanuma H., Ohno N., Toda H., (2004), A study of adhesive strength of cold spray coatings, *ITSCE 2004*, Proceeding (CD), Düsseldorf.

60 Gabel H., Taphorn R.M., (1997), Solid-state spray forming of aluminum near-net shapes, *JOM*, 8, pp. 31–33.

61 Gabel, H. (2004), Kinetic metallization compared with HVOF *Adv. Mater. Process.*, 162(5), pp. 47–48.

62 Gärtner, F., Borchers, C., Stoltenhoff, T., Kreye, H., Assadi, H.(2003), Numerical and microstructural investigations of the bonding mechanisms in cold spraying, *Thermal Spray 2003: Advancing the Science & Applying the Technology*, 1, pp. 1–8.

63 Gerland M. Presles H.N., Guin J.P., Bertheau D. (2000), Explosive cladding of a thin Ni-film to an aluminium alloy *Mater. Sci. Eng. A (Struct. Mater., Prop., Microstruct. Process.)*, A280, pp. 311–319.

64 German, R.M., (1985), *Liquid Phase Sintering*, Plenum Press, New York.

65 Gidaspow D., (1994), *Multiphase Flow and Fluidization: Continuum and Kinetic Theory Descriptions*, Academic Press, Boston.

66 Gilmore D.L., Dykhuizen R.C., Neiser R.A., Roemer, Smith M.F., (1999), Particle velocity and deposition efficiency of cold spray process, *J. Therm. Spr. Technol.*, 8(40), pp. 576–582.

67 Gotoh M, Yamashita M, (2001), A study of high rate shearing of commercially pure aluminum sheet, *J. Mater. Process. Technol.*, 110, pp. 253–264.

68 Graham R.A, Thadhani N.N, (1993), in: A.B. Sawaoka (Ed.), *Shock Waves in Materials Science*, Springer, Berlin, pp. 35–65.

69 Green D.J. (1998), *An Introduction to the Mechanical Properties of Ceramics*, Cambridge University Press, Cambridge, p. 336.

70 Griffith A.A, (1920), The phenomenon of rupture and flow in solids, *Phil. Trans. Roy: Soc., Lond.*, A221, pp. 163–98.

71 Grodzinski A., Senkara J., Kozlowski, M., (1996), Duralumin-stainless steel joint: structure and mechanism of forming by means of electron beam welding and Ag–2Mg filler metal shim, *J. Mater. Sci.*, 31(18), pp. 4967–4973.

72 Grujicic M., Saylor J.R., Beasley D.E., DeRosset W.S., Helfrich D., (2003), Computational analysis of the interfacial bonding between feed-powder particles and the substrate in the cold-gas dynamic-spray process, *Appl. Surf. Sci.*, 219 , pp. 211–227.

73 Grujicic M., Zhao C.L., DeRosset W.S., Helfritch D., (2004), Adiabatic shear instability based mechanism for particles/substrate bonding in the cold-gas dynamic-spray process, *Mater. Des.*, 25, pp. 681–688.

74 Hammerschmidt M., Kreye H., (1981), Microstructure and bonding mechanism in explosive welding, in: M.A. Meyers, L.E. Murr (Eds.), *Shock Waves and High Strain-Rate Phenomena in Metals*, Plenum Press, New York, pp. 961–973.

75 Helfritch D., Champagne V., (2006), Optimal particle size for the cold spray process, *Proceeding of the ITSCE*, May 15–18, 2006, Seattle, Washington, USA, ASM International, Materials Park, OH.

76 Helle A.S., Easterling K.E., Ashby M.F., (1985), Hot isostatic pressing diagrams: new developments, *Acta Metallurgica*, 33(12), pp. 2163–2174.

77 Henderson C.B., (1976), Drag coefficients of spheres in continuum and rarefied flows, *AIAA J.*, 14, pp. 707–708.

78 Hinze J.O., (1975), *Turbulence*, McGraw-Hill, New York.

79 Hu N., Molinari J.F., (2004), Shear bands in dense metallic granular materials, *J. Mech. Phys. Solids*, 52, pp. 499–531.

80 Irwin G.R., (1998), Fracture, in: *Handbuch der Physik*, Vol. 6, Springer, Berlin, pp. 551–89.

81 Jen T.C., Li L., Cui W., Chen Q., Zhang X., (2005), Numerical investigations on cold gas dynamic spray process with nano- and microsize particles, *Int. J. Heat Mass Transf.*, 48, pp. 4384–4396.

82 Jeng J.Y., Peng S.C., Chou C.J., (2000), Metal rapid prototype fabrication using selective laser cladding technology, *Int. J. Adv. Manuf. Technol.* 16, pp. 681–687.

83 Jenkins J.T., (1992), Boundary conditions for rapid granular flow: Flat, frictional walls, *J. Appl. Mech.*, 59, pp. 120–134.

84 Jodoin B., Raletz F., Vardelle M., (2006), Cold spray modeling and validation using an optical diagnostic method, *Surf. Coat. Technol.*, 200, pp. 4424–4432.

85 Johnson G.R., Cook W.H., (1983), A constitutive model and data for metals subjected to large strains, high strain rates, and high temperatures. in: *Proceedings of the 7th International Symposium on Ballistrics,* Organized under the auspices of the Royal Institution of Engineers (Klvl), Division for Military Engineering in cooperation with the American Defense Preparedness Association, The Hague, pp. 541–547.

86 Jones, D.A., (1992), *Principles and Prevention of Corrosion*, Maxwell Macmillan International Editions, New York.

87 Kalla G., (1996), Soudage au laser CO_2 de produits en acier de construction pouvant avoir jusqu'a 20 mm d'epaisseur (CO_2-laser beam welding of structural steel with a thickness up to 20 mm) *Revue de Metallurgie. Cahiers D'Informations Techniques*, 93(10), p. 1303.

88 Kannatey-Asibu Ir E., (1997), Milestone developments in welding and joining processes, *Journ. Manufacting Science and Engineering*, ASME 119(4), pp. 801–810.

89 Karthikeyan J. (2005), The cold spray process has the potential to reduce costs and improve quality in both coatings and freeform fabrication of near-net-shape parts, *Adv. Mater. Process.*, (3), p. 35.

90 Karthikeyan J., Kay C.M., Lindeman J., Lima R.S., Berndt C.C., (2000), in: C.C. Berndt (Ed.), *Thermal Spray: Surface Engineering via Applied Research*, ASM International, OH, pp. 255–262.

91 Karthikeyan J., Kay C.M., Lindemann, Lima R.S., Berndt C.C., (2001), in: C.C. Berndt, K.A. Khor, E:E. Lugscheider (Eds.), *Thermal Spray J. 2001: New Surfaces for a New Millenium*, ASM International, OH, pp. 383–387.

92 Karthikeyan, J., Laha, T., Balani, K., Agarwal, A, Munroe N., (2004), Microstructural and electrochemical characterization of cold-sprayed 1100 aluminum coating, *ITSCE 2004*, Proceeding (CD), Düsseldorf.

93 Kashirin A.I., Klyuev O.F., Buzdygar T.V., (2002), Apparatus for gas-dynamic coating, US Patent 6,402,050.

94 Kay A., Karthikeyan J., (2003), Advanced cold spray system, US Patent 6,502,767.

95 Kay A., Karthikeyan J., (2004), Cold spray system nozzle, US Patent 6,722,584.

96 Kiselev S.P., Kiselev V.P., (2002), Superdeep penetration of particles into a metal target, *Int. J. Impact Eng.* 27, pp. 135–152.

97 Kitron A., Elperin T., Tamir A., (1981), Monte Carlo analysis of wall erosion and direct contact heat transfer by impinging two-phase jets, *J. Thermophys.*, 3, pp. 112–122.

98 Klepaczko J.R., (1988), A general approach to rate sensitivity and constitutive modeling of FCC and BCC metals, in: A.A. Balkema (Ed.), *Impact: Effects of Fast Transient Loadings*, Rotterdam, Balkema Publishers, pp. 3–17.

99 Klinkov S.V., Kosarev V.F., Rein M., (2005), Cold spray deposition: significance of particle impact phenomena, *Aerosp. Sci. Technol.*, 9, pp. 582–591.

100 Kochs U.F., Argon A.S., Ashby M.F., (1975), *Thermodynamics and Kinetics of Slip*, Pergamon Press, Oxford, UK.

101 Kosarev V.F., Klinkov S.V., Alkhimov A.P., Papyrin A.N., (2003), On some aspects of gas dynamics of the cold spray process, *J. Therm. Spray Technol.*, 12(2), pp. 265–281.

102 Kouzeli M., Mortensen A., (2002), Size dependent strengthening in particle reinforced aluminium, *Acta Mater.*, 50, pp. 39–51.

103 Krautkramer J., Krautkramer H., (1983), *Ultrasonic Testing of Materials*, 3rd edn., Springer-Verlag, Berlin, pp. 580–587.

104 Kreye H., (2004), The cold spray process and its potential for industrial applications, *Cold Spray* (CD Proceeding), Akron OH, September 2004.

105 Kreye H., Stoltenhoff T., (2000), in: C.C. Berndt (Ed.), *Thermal Spray: Surface Engineering via Applied Research*, ASM International, OH, pp. 419–422.

106 Kussin J., Sommerfeld M., (2002), Experimental studies on particle behaviour and turbulence modification in horizontal channel flow with different wall roughness, *Exp. Fluids*, 33, pp. 143–159.

107 Lall C., Heath W., (2000), P/M aluminium structural parts – manufacturing and metallurgical fundamentals, *Int. J. Powder Met.*, 36(6), pp. 45–50.

108 Lee D.N., Kim H.S., (1992), Plastic yield behavior of porous metals, *Powder Metallurgy*, 35(4), pp. 275–279.

109 Leshchynsky V., (1989), Effect of strain hardening during shear of powder compact, *Experiments Bulletin of High Education Schools, Iron Metallurgy* (Izvesia VUZov in Russian), pp. 103–105.

110 Leshchynsky V., Strumban E., Maev R.Gr., (2005), Characterization of powder mixtures for gas dynamic spraying (GDS) by shear compression, *Montreal PM^{2}TEC Conference Proceedings*, (CD Proceeding).

111 Levitas V.I., Nesterenko V.F., Meyers M.A., (2000), Strain-induced structural changes and chemical reactions – I. Thermomechanical and kinetic models, *Acta Metall.*, 46 (16), pp. 5929–5945.

112 Li X., Dunn P.F, Brach R.M., (1999), Experimental and numerical studies on the normal impact of microspheres with surfaces, *J. Aerosol Sci*, 30(4), pp. 439–449.

113 Li C.J., Li W.Y., (2003), Deposition characteristics of titanium coating in cold spraying, *Surf. Coat. Technol.*, 167, pp. 278–283.

114 Li C.J. Ohmori A., Harada Y., (1996), Formation of an amorphous phase in thermally sprayed WC–Co, *Surf. Coat. Technol. J. Thermal Spray Technol.*, 5(1), pp. 69–73.

115 Li X.C., Stampfl J., Fritz B.P., (2000), Mechanical and thermal expansion behavior of laser deposited metal matrix composites of Invar and TiC, *Mater. Sci. Eng. A* 282, p. 86.

116 Li C.J., Yang G.J., Huang X.C., Li W.Y., (2004), Formation of TiO_2 photocatalyst through cold spraying, *Thermal Spray 2004: Advances in Technology and Application*, ASM International, May 10–12, 2004 (Osaka, Japan), ASM International, pp. 315–319.

117 Li J.R., Yu J.L., Wei Z.G., (2003), Influence of specimen geometry on adiabatic shear instability of tungsten heavy alloys, *Int. J. Impact Eng.*, 28, pp. 303–314.

118 Li Z., Zhu J., Zhang C., (2005), Numerical simulations of ultrafine powder coating systems, *Powder Technol.*, 150, pp. 155–167.

119 Lima R.S., Karthikeyan J., Kay C.M., Lindemann J., Berndt C.C., (2002), Microstructural characteristics of cold-sprayed nanostructured WC-Co coatings, *Thin Solid Films*, 416(1–2), pp. 129–135.

120 Louge, M.Y., Mastorakos E., Jenkins J.T., (1991), The role of particle collisions in pneumatic transport, *Journ. Fluid Mechanics*, 231, pp. 345–359.

121 MacKenzie J.K., (1950), The elastic constants of solid spherical holes, *Proc. Phys. Soc. London, B*, 63, pp. 2–11.

122 Maev R.Gr., (2007), *Acoustic Microscopy*, Wiley-VCH Verlag GmbH, Weinheim, in press.

123 Maev R.Gr., Leshchynsky E.V., (2005), Compaction and bonding alloyed powder by air gas dynamic spray technology, *Montreal PM^{2}TEC Conference Proceedings*, (CD Proceeding).

124 Maev R.Gr., Leshchynsky E.V., (2006a), Air gas dynamic spraying of powder mixtures: theory and application, *J. Therm. Spray Technol.*, pp. 198–205.

125 Maev R.Gr., Leshchynsky E.V., (2006b), Compaction and bonding alloyed powder by air gas dynamic spray technology, *Surf. Coat. Technol.*, (to be submitted).

126 Maev R.Gr., Leshchynsky E.V., (2006c), Low pressure gas dynamic spray: shear localization during particle shock consolidation, *International Thermal Spray Conference Proceedings*, Seattle, USA, (CD Proceeding.)

127 Maev R.Gr., Leshchynsky E.V., Papyrin A., (2006), Structure formation of Ni-based composite coatings during low pressure gas dynamic spraying, *International Thermal Spray Conference Proceedings*, Seattle, USA , (CD Proceeding).

128 Maev R.Gr., O'Neill B.E., Aczel A., Leshchynsky E.V., (2005), Acoustic characterization of powder coating adhesion, *Prague EUROPM Congress Proceedings*, (CD Proceeding).

129 Maev R.Gr., Salloum G., Islam M.U., Aczel A., Leshchynsky V., (2005), Air gas dynamic spraying of powder mixtures: theory and application, *ITSCE Proceedings* (CD), Basel, Switzerland, pp. 1121–1126.

130 Maev R.Gr., Strumban E., Leshchynsky V., Beneteau M., (2004), Supersonic induced mechanical alloy technology and coatings for automotive and Aerospace Applications, *Cold Spray 2004: An Emerging Spray Coating Technology*, September 27–28, 2004 (Akron, Ohio), ASM International, Electronic.

131 Maeva E., Aczel A., Leshchynsky E.V., (2006), Adiabatic shear band formation in the gas dynamic spray process, *J. Mater. Charact.* (submitted).

132 Marchand A., Duffy J., (1988), An experimental study of the formation process of adiabatic shear bands in a structural steel, *J. Mech. Phys. Solids*, 36(1), p. 25.

133 Marrocco T., McCartney D.G., Shipway P.H., (2005), Production of titanium deposits by cold gas dynamic spray, *ITSCE 2005*, Proceeding (CD), Basel, Switzerland.

134 Martin, L.P., Rosen M., (1997), Correlation between surface area reduction and ultrasonic velocity in sintered zinc oxide powders, *J. Am. Ceram. Soc.*, 80, pp. 839–846.

135 Mazumder J., Choi J., Nagarathnam K., Koch J., Hetzner D., (1997), *J. Miner. Metals Mater. Soc.*, 49, 55.

136 McCune R.C., (2003), Potential applications of cold spray technologies in automotive manufacturing, in: C. Moreau, B. Marple (Eds.), *Thermal Spray 2003: Advancing the Science and Applying the Technology*, ASM International, Materials Park, OH, pp. 63–70.

137 McCune R.C., Donlon W.T., Cartwright E.L., Papyrin A.N., Rybicki E.E., Shadley J.R., (1996), in: C.C. Berndt (Ed.), *Thermal Spray: Practical Solutions for Engineering Problems*, ASM International, Materials Pack, OH, pp. 397–403.

138 McCune R.C., Donlon W.T., Popoola O.O., Cartwright E.L., (2000), Characterization of copper layers produced by cold gas-dynamic spraying, *J. Thermal Spray Technol.*, 9(1), pp. 73–81.

139 McCune R.C., Papyrin A.N., All J.N., Riggs W.L., Zajchowski P.H., (1995), An exploration of the cold gas dynamic spray method for several materials systems, in: *Proceedings of the 8th National*

Thermal Spray Conference, Houston, Texas, pp. 1–5.

140 Mebtoul M., Large J., Guidon P., (1996), High velocity impact of particles on a target – an experimental study, *J. Miner. Process.*, 44–45, pp. 77–91.

141 Meyers M.A., Benson D.J., Olevsky E.A., (1999), Shock consolidation: microstructurally-based analysis and computational modeling, *Acta Mater.*, 47(7), pp. 2089–2108.

142 Meyers M.A, Nesterenko V.E., LaSalvia J.C., Xue Q., (2001), Shear localization in dynamic deformation of materials: microstructural evolution and self-organization, *Mater. Sci. Eng.* pp. 204–225.

143 Milewski J.O., Lewis G.K., Thoma D.J., Keel G.I., Nemec R.B., Reinert R.A., (1998), Directed light fabrication of a solid metal hemisphere using 5-axis powder deposition, *J. Mater. Process. Technol.*, 75 , 165

144 Mizusaki J., Tsuchiya S., Waragai K., Tagawa H., Arai Y., Kuwayama Y., (1996), Simple mathematical model for the electrical conductivity of highly porous ceramics, *Journal of the American Ceramic Society*, 79(1), Jan., pp. 109–122.

145 Mohanty M., Smith R.W., (1996), Sliding wear behavior of thermally sprayed 75/25Cr3C2/NiCr wear-resistant coatings, *Wear*, 198, pp. 251–266.

146 Molinari A., (1997), Collective behaviour and spacing of adiabatic shear bands, *J. Mech. Phys. Solids*, 45(9), pp. 1551–1575.

147 Molinari J.F., Ortiz M., (2002), A study of solid-particle erosion of metallic targets, *Int. J. Impact Eng.*, 27 , pp. 347–358.

148 Morgan R., Fox P., Pattison J., Sutcliffe C., O'Neill W., (2004), Analysis of cold gas dynamically sprayed aluminium deposits, *Mater. Lett.*, 58, pp. 1317–1320.

149 Muehlberger E., de la Vega. P., (1989), Powder feeder, US Patent 4,808,042.

150 Mukhopadhyay A.K,. Phani K.K, (1998), Young's modulus–porosity relations: an analysis based on a minimum contact area model, *J. Mat. Sci.*, 33, pp. 69–72.

151 Mukhopadhyay A.K., Phani K.K., (2000), An analysis of microstructural parameters in the minimum contact area model for ultrasonic velocity-porosity relations, *J. Am. Ceram. Soc.*, 20, pp. 29–38.

152 Naroyana Murty S.V.S., Nageswara Rao B., Kashyap B.P., (2000), Instability criteria for hot deformation of materials, *Int. Mater. Rev.*, 45(1), pp. 15–26.

153 Narayanasamy R., Ramesh T., Pandey K.S., (2005), Workability studies on cold upsetting of Al–Al_2O_3 composite material, *Materials and Design*, 27(7), pp. 566–575.

154 Neiser R.A., Brockman J.E., Ohern T.J., Smith M.F., Dykhuizen R.C., Roemer T.J., Teets R.E., (1995), Wire melting and droplet atomization in a high velocity oxy-fuel jet, in: C.C. Berndt, S. Sampath (Eds.), *Thermal Spray Science and Technology*, ASM International, Materials Park, OH, pp. 99–104.

155 Nesterenko, V.F , (1995), Dynamic loading of porous materials: potential and restrictions for novel materials applications, in: L.E. Murr, K.P. Staudhammer, M.A. Meyers (Eds.), *Metallurgical and Materials Applications of Shock-Wave and High-Strain-Rate Phenomena. Proceedings of the 1995 International Conference*, 6–10 August 1995 , El Paso, TX, USA, Elsevier, Amsterdam, pp. 3–13.

156 Niranatlumpong P., Koiprasert H., (2006), Improved corrosion resistance of thermally sprayed coating via surface grinding and electroplating techniques, *Surf. Coat. Technol.*, 201, pp. 737–743.

157 Oesterle B., Petitjean A., (1993), Simulation of particle-to-particle interactions in gas–solid flows, *Int. J. Multiph. Flow*, 19, pp. 199–211.

158 Olevsky, E.A. Boccaccini, A.R., Wilson, (1998), S.D.R. Comments and author's reply on "The measurement of bulk viscosity and the elastic-viscous analogy," *Acta Mech.*, 128(3–4), pp. 263–266.

159 Orrhede M., Tolani R., Salama K., (1996), Elastic constants and thermal expansion of aluminium–SiC metal-matrix composites, *Res. Nondestr. Eval.* 8, pp. 23–37.

160 Papyrin A., (2001), Cold spray technology, *Adv. Mater. Process.*, 159(9), pp. 49–51.

161 Papyrin A.N., Kosarev V.F., Klinkov S.V., Alkhimov A.P. (2002), On the interaction of high speed particles with a substrate under cold spraying, in: E.F. Lugscheider, C.C. Berndt (Eds.), *Proceeding of the International Thermal Spray Conference*, 2002, Dusseldorf, Germany, DVS-Verlag, pp. 380–386.

162 Papyrin A.N., Kosarev V.F., Klinkov S.V., Alkhimov A.P., Fomin V. M. (2006), *Cold Spray Technology*, Elsevier, Amsterdam, ISBN-10: 0080451551, p. 336.

163 Peng H.X., Fan Z., Evans J.R.G., (2001), Bi-continuous metal matrix composites, *Mater. Sci. Eng. A* 303, pp. 37–45.

164 Procopio A.T., Zavaliangos A., Cunningham J.C., (2003), Analysis of the diametrical compression test and the applicability to plastically deforming materials, *J. Mater. Sci.*, 38, pp. 3629–3639.

165 Prummer R., (1987), Explosivverdichting Pulvriger Substanzen Grundlagen Verfahrenergebnisse, Springer, Berlin, (in German).

166 Raghukandan K., (2003), Analysis of the explosive cladding of Cu-low carbon steel plates, *J. Mat. Proc. Technol.*, 139, pp. 573–577.

167 Rajaratnam N., (1976), *Turbulent Jets*, Elsevier, Amsterdam.

168 Raletz F., Ezo'o G., Vardelle M., Ducos M., (2003), Characterization of cold-sprayed nickel-base coatings, in: B.R. Marple, C. Moreau (Eds.), *Thermal Spray 2003: Advancing the Science & Applying the Technology*, Orlando, FL, ASM International, Materials Park, OH, pp. 45–50.

169 Ranz W.E., Marshall J.R.W.R., (1952), Evaporation from drops, *Chem. Eng. Prog.*, 48, pp. 173–180.

170 Raybould D., (1981), The properties of stainless steel compacted dynamically to produce cold interparticle welding, *J. Mater. Sci.*, 16(3), pp. 589–598.

171 Rice R.W., (1996), Evaluating porosity parameters for porosity-property relations, *J. Am. Ceram. Soc.*, 76, pp. 1801–1805.

172 Rocheville C.F., (1963), US Patent 3,100,724, August 13.

173 Rosato A.D., Vreeland T.Jr., Prinz F.B., (1991), Manufacture of powder compacts, *Int. Mater. Rev.*, 36(2), pp. 45–61.

174 Rothbart H.A., (Ed.), (1996), *Mechanical Design Handbook*, McGraw–Hill, New York, p. 94.

175 Rotolico A.J., Romero E., Lyons J.E., (1983), Device for the controlled feeding of powder material, US Patent 4,381,898.

176 Rudinger G., (1980), Fundamentals of gas-particle flow, *Handbook of Powder Technology*, Vol. 2, Elsevier, Amsterdam.

177 Sabersky R.H., Acosta A.J., Hauptmenn E.G., (1971), *Fluid Flow*, Macmillan, London.

178 Sakaki K., Shimizi Y., (2001), Effect of the increase in the entrance convergent section length of the gun nozzle on the high-velocity oxygen fuel and cold spray process, *J. Thermal Spray Technol.*, 10(3), pp. 487–496.

179 Sampath S., Jiang X.Y., Matejicek J., Prchlik L., Kulkarni A., Vaidya A. (2004), Role of thermal spray processing method on the microstructure, residual stress and properties of coatings: an integrated study for Ni-5 wt.%Al bond coats , *Mater. Sci. Eng.*, A364, pp. 216–231.

180 Savitskii A.P., (1993), *Liquid Phase Sintering of the Systems with Interacting Components*, Science, Tomsk, Russia.

181 Schinella A.A., (1971), Powder feeder, US Patent 3,618,828.

182 Schlichting H., (1979), *Boundary Layer Theory*, McGraw-Hill, New York.

183 Schmidt J, Dorner H., Tenckhoff E., (1990), Manufacture of complex parts by shape welding, *J. Nuclear Mater.*, 171(1), pp. 120–127.

184 Schmidt T., Gartner F., Assadi H., Kreye H., (2006), Development of a generalized parameter window for cold spray deposition, *Acta Mater.*, 54, pp. 729–742.

185 Schmidt T., Gartner F., Kreye H. (2003), High strain rate deformation phenomena in explosive powder compaction and cold gas spraying, in: B.R. Marple, C. Moreau, (Eds.), *Proceedings of the ITSC 2003*, Orlando, ASM International, Materials Park, OH, pp. 9–17.

186 Schoenfeld E., Wright T.W., (2003), A failure criterion based on material instability, *Int. J. Solids Struct.*, 40(4) , pp. 3021–37.

187 Schwarz R.B., Kasiraj P., Vreeland T. Jr., Ahrens T.J., (1984), Theory for the shock wave consolidation of poweders. *Acta Metall.*, 32, p. 1243.

188 Segal V.M., (2002), Severe plastic deformation: simple shear versus pure shear, *Mater. Sci. Eng.*, A338, pp. 331–344.

189 Senior R.C., Grace J.R., (1998), Integrated particle collision and turbulent diffusion model for dilute gas–solid suspensions, *Powd. Technol.*, 96(1), pp. 48–78.

190 Seong-Jun Park, Heung Nam Han, Kyu Hwan Oh, Dong Nyung Lee, (1999), Model of compaction of metal powders, *Int. J. Mech. Sci.*, 41, pp. 121–141.

191 Shapiro A.H., (1953), *The Dynamics and Thermodynamics of Compressible Fluid Flow*, Ronald Press, New York.

192 Simchi A., Veltl G., (2003), Investigation of warm compaction and sintering behaviour of aluminium alloys, *Powder Met.*, 46(2), pp. 159–164.

193 Simmons J.W., Wilson R.D., (1996), Joining of high-nitrogen stainless steel by capacitor discharge welding, *Weld. J. (Miami, Fla)*, 75(6), pp. 185–190.

194 Smith, M.F., Brockmann, J.E., Dykhuizen, R.C., Gilmore, D.L., Neiser, R.A., Roemer, T.J., (1999), Cold spray direct fabrication – high rate, solid state, material consolidation, *Materials Research Society Symposium Proceedings*, 542, pp. 65–76.

195 Smith M.F., Dykhuizen R.C., (1988), The effect of chamber pressure on particle velocities in low pressure plasma spray deposition, *Surf. Coat. Technol.*, 34(1), pp. 25–31.

196 Sommerfeld M., (2003), Analysis of collision effects for turbulent gas-particle flow in a horizontal channel: Part I. Particle transport, *Int. J. Multiph. Flow*, 29, pp. 675–699.

197 Sommerfeld M., Kussin J., (2003), Analysis of collision effects for turbulent gas-particle flow in a horizontal channel: Part II. Inegral properties and validation, *Int. J. Multiph. Flow*, 29, pp. 701–718.

198 Stoltenhoff T., Kreye H., Krommer W., Richter H.J., (2000), Cold spraying-from thermal spraying to high kinetic energy spraying, in: P. Heinrich (Ed.), *HVOF Colloquium 2000*, Gemeinschaft Thermisches Spritzen e.V., 2000, pp. 29–38.

199 Stoltenhoff T., Kreye H., Richter H.J., Assadi H., (2001), Optimization of the cold spray process, in: C.C Berndt, K.A. Khor, E.E. Lugscheider (Eds.), *Thermal Spray (2001): New Surfaces for a New Millenium*, ASM International, Materials Park, OH, pp. 409–416.

200 Stoltenhoff T., Kreye H., Richter H.J., (2002), An analysis of the cold spray process and its coating, *J. Thermal Spray Technol.* 11(4), pp. 542–550.

201 Straffelini G., Pellizzari M., Molinari A., (2004), Influence of load and temperature on the dry sliding behavior of Al-based metal matrix-composites, *Wear*, 256, pp. 754–763.

202 Streeter V.L., (1961), *Handbook of Fluid Dynamics*, McGraw-Hill, New York.

203 Suegama P.H., Espallargas N., Guilemany J.M., Fernández J., Benedettia A.V, (2006), Electrochemical and structural characterization of heat-treated Cr3C2–NiCr coatings, *J. Electrochem. Soc.*, 153(10), pp. B434–B445.

204 Suh N.P., (1986), *Tribophysics*, Prentice-Hall, New Jersey, pp. 63–90.

205 Sun Z., Karppi R., (1996), Application of electron beam welding for the joining of dissimilar metals: an overview, *J. Mater. Process. Technol.*, 59, pp. 257–267.

206 Tang F., Anderson I.E., Biner S.B., (2003), Microstructures and mechanical properties of pure Al matrix composites reinforced by Al–Cu–Fe alloy particles, *Mater. Sci. Eng. A* , 363, pp. 20–29.

207 Tanaka T., Tsuji Y., (1991), Numerical simulation of gas-phase flow in a vertical pipe: on the effect of interparticle collision, in: D.E. Stock et al. (Eds.), *Gas–Solid Flows*, ASME FED, Vol. 121, pp. 123–128.

208 Tanga F., Meeks H., Spowart J.E., Gnaeupel-Herold T., Prask H., Anderson I.E., (2004), Consolidation effects on tensile properties of an elemental Al matrix composite, *Mater. Sci. Eng. A* , 386, pp. 194–204.

209 Tapphorn R.M., Gabel H., (2004), Powder fluidizing devices and portable powder-deposition apparatus for coating and spray forming, US Patent 6,715,640.

210 Terheci M., Manory R., Hensler J., (1995), The friction and wear of automotive grey cast iron under dry sliding conditions, *Wear*, 185, pp. 119–124.

211 Tjong S.C., Zhu S.M., Ho N.J., Ku J.S., (1995), Microstructural characteristics and creep rupture behavior of electron beam and laser welded AISI 316L stainless steel, *J. Nuclear Mater.*, 227(1–2), p. 24.

212 Tokarev A.O., (1996), Structure of aluminum powder coatings prepared by cold gas dynamic spraying, *Met. Sci. Heat Treat.*, 35, p. 136.

213 Tong W., Ravichandran G., Christman T., Vreeland T. Jr. (2003), Processing SiC-particulate reinforced titanium-based metal matrix composites by shock wave consolidation, *Acta Metall. Mater.*, 43(1), pp. 235–250.

214 Triesch O., Bohnet M., (2001), Measurement and CFD prediction of velocity and concentration profiles in a decelerated gas–solid flow, *Powder Technol.* 115, pp. 101–113.

215 Tsirkunov Yu.M., (2001), Gas-particle flows around bodies – key problems, modeling and numerical analysis, in: E. Michaelides (Ed.), *Proc. 4th Int. Conf. Multiphase Flow*, New Orleans, USA (CD-Rom Proc. ICMF' 2001, paper 609).

216 Tsuji Y., Morikawa Y., and Schiomi H., (1984), LDV measurements of an air–solid two-phase flow un a vertical pipe, *J. Fluid Mech.*, 139, p. 417.

217 Tszeng T.C., Wu W.T., (1996), A study of the coefficients in yield functions modeling metal powder deformation, *Acta Mater.*, 44(9), pp. 3543–3552.

218 Turgutlu A., Al-Hassani S.T.S., Akyurt M., (1995), Experimental investigation of deformation and jetting during impact spot welding, *Int. J. Impact Eng.*, 16, (5–6), p. 789.

219 Turgutlu A., Al-Hassani S.T.S, Akyurt, M., (1997), Assessment of bond interface in impact spot welding, *Int. J. Impact Eng.*, 19(9–10), pp. 755–767.

220 Underwood E.E. (1985), *Metals Handbook*, 9th edn., ASM International, Metals Park (OH), pp. 123–34.

221 Van Steenkiste T.H., (2003), Method of producing a coating using a kinetic spray process with large particles and nozzles for the same, US Patent 6,623,796.

222 Van Steenkiste T.H., (2004), Low pressure powder injection method and system for a kinetic spray process, US Patent No 6,811,812.

223 Van Steenkiste T.H., Smith J.R., (2003), Evaluation of coatings produced via kinetic and cold spray processes, in: *Proceedings of the International Thermal Spray Conference*, ASM International, Materials Park, OH, USA, Orlando, 5–8 May, pp. 53–61.

224 Van Steenkiste T.H., Smith J.R., Teets R.E., (2002), Aluminum coatings via kinetic spray with relatively large powder particles, *Surf. Coat. Technol.*, 154, pp. 237–252.

225 Van Steenkiste T.H., Smith J.R., Teets R.E., Moleski J.J., Gorkiewicz D.W., Tison R.P., Marantz D.R., Kowalsky K.A., Riggs W.L., Zajchowski P.H., Pilsner B., McCune R.C., Barnett K.J., (1999), Kinetic spray coatings, *Surf. Coat. Technol.*, 111, pp. 62–71.

226 Villafuerte J., (2005), Cold spray: A new technology, *Weld. J.*, 84(5), pp. 24–29.

227 Vlcek J., Gimeno L., Huber H., Lugscheider E., (2003), A systematic approach to material eligibility for the cold spray process, *Thermal Spray 2003: Advancing the Science & Applying the Technology*, B.R. Marple, C. Moreau (Eds.), May 5–8, Orlando, FL, ASM International, Materials Park, OH, pp. 37–44.

228 Volkov A., Tsirkunov Yu., Oesterle B., (2005), Numerical simulation of a supersonic gas–solid flow over a plunt body: The role of inter-particle collisions and two-way coupling effects, *Int. J. Multiph. Flow*, 31, pp. 1244–1275.

229 Voyer J., Stoltenhoff T., Kreye H., (2003), Development of cold gas sprayed coatings, in: *Proceedings of the International Thermal Spray Conference*, 5–8 May, ASM International, Materials Park, OH, USA, Orlando, pp. 71–78.

230 Wallis G.B. (1969), *One-Dimensional Two-Phase-Flow*, McGraw-Hill, New York.

231 Walsh J.M., (1975), Drag coefficient equation for small particles in high speed flows, *AIAA J.*, 13(11), pp. 1526–28.

232 Walter J.W., (1992), Numerical experiments on adiabatic shear band formation in one dimension, *Int. J. Plast.*, 8, pp. 657–693.

233 Wang H.M., Chen Y.L., Yu L.G., (2000), "Insitu" weld alloying/laser beam welding of SiCp/6061Al MMC, *Mater. Sc. and Eng. A*, 293(1–2), pp. 1–6.

234 Weast R.C., (2000), *Handbook of Physics and Chemistry*, 81st edn., CRC Press, Boca Raton, FL.

235 Weinert H., Maeva E., Leshchynsky V., Low pressure gas dynamic spray forming near-net shape parts, *International Thermal Spray Conference 2006 Proceedings*, Seattle, USA, (CD Proceeding).

236 Williams J.A., (1996), Analytical models of scratch hardness, *Tribol. Int.*, 29(8), pp. 675–694.

237 Wright T.W., (1987), Steady shearing in a viscoplastic solid, *J. Mech. Phys. Solids*, 35, p. 269–282.

238 Wright T.W., (1992), Shear band susceptibility: work hardening materials, *Int. J. Plast.*, 8(2), 583–602.

239 Wright T.W., (2002), *The Physics and Mathematics of Adiabatic Shear Bands*, Cambridge University Press, Cambridge, New York , p. 241.

240 Wright T.W., Batra R.C., (1985), The initiation and growth of adiabatic shear bands, *Int. J. Plast.*, 1, pp. 205–212.

241 Wu J., Fang H., Lee C., Yoon S., Kim H., (2006), Critical and maximum velocities in kinetic spraying, *Proceedings of the 2006 International Thermal Spray Conference*, Seattle, Washington, USA.

242 Wu J., Fang H., Yoon S., Kim H., Lee C., (2005), Measurement of particle velocity and characterization of deposition in aluminium alloy kinetic spraying process, *Appl. Surf. Sci. 252*, pp. 1368–1377.

243 Yilbas B.S., Sami M., Nickel J., Coban A., Said S.A.M., (1998), Introduction into the electron beam welding of austenitic 321-type stainless steel, *J. Mater. Process. Technol.*, 82, p. 13.

244 Xu B., Zhu Z., Ma S., Zhang W., Liu W., (2004), Sliding wear behavior of Fe–Al and Fe–Al/WC coatings prepared by high velocity arc spraying, *Wear*, 257, pp. 1089–1095.

245 Zerilli J., Armstrong R.W., (1997), Dislocation mechanics based analysis of material dynamics behaviour: enhanced ductility, deformation twinning, shock deformation, shear instability, dynamic recovery, *J. Phys.* 7, pp. 637–642.

246 Zhang D.L., (2004), Processing of advanced materials using high-energy mechanical milling progress, *Mat. Sci.*, 49, pp. 537–560.

247 Zhang D., Shipway P.H., McCartney D.G., (2003a), Particle-substrate interactions in cold gas dynamic spraying, in: B.R. Marple, C. Moreau (Eds.), *Thermal Spray 2003: Advancing the Science & Applying the Technology*, May 5–8, 2003, Orlando, FL, ASM International, pp. 45–52.

248 Zhang H., Wang G., Luo Y., Nakaga T., (2001), Rapid hard tooling by plasma spraying for injection molding and sheet metal forming, *Thin Solid Films*, 390, p. 7.

249 Zhang H., Xu J., Wang G., (2003b), Fundamental study on plasma deposition manufacturing, *Surf. Coat. Technol.*, 171, pp. 112–118.

250 Zhou F., Wright T.W., Ramesh K.T., (2006), A numerical methodology for investigating the formation of adiabatic shear bands, *J. Mech. Phys. Solids* 54 , pp. 904–926.

251 Zimmerly C.A., Inal O.T., Richman R.H., (1994), Explosive welding of a near-equiatomic nickel-titanium alloy to low-carbon steel, *Mater. Sci. Eng. A, Struct. Mater., Prop. Microstruct. Process. A*, 188, (1–2), 30. pp. 251–2514.

252 Zukas J.A., (1990), *High-Velocity Impact Dynamics*, Wiley, New York.

Index

Introduction to Low Pressure Gas Dynamic Spray Roman Gr. Maev and Volf Leshchynsky

ISBN: 978-3-527-40659-3